THE TROPICAL AGRICULTURALIST

Series Editor
René Coste

Formerly President of the IRCC

General Editor, livestock volumes
Anthony J. Smith

Centre for Tropical Veterinary Medicine
University of Edinburgh

Animal Health

Volume 1   General Principles

Archie Hunter

Centre for Tropical Veterinary Medicine
University of Edinburgh

With the collaboration of
Professor Uilenberg
Institut d'Elevage et de Médecine Vétérinaire
des Pays Tropicaux
Maisons-Alfort
France

This edition first published 1996

The Tropical Agriculturalist Series originated under
the title *Le Technicien d'Agriculture Tropicale* published
by G. P. Maisonneuve et Larose, 15 rue Victor-Cousin,
75005 Paris, France, in association with the Agency
for Cultural and Technical Co-operation based in Paris,
France. Volumes in the series in the French language
are available from Maisonneuve et Larose.

Published by MACMILLAN EDUCATION LTD
London and Basingstoke
*Associated companies and representatives in Accra, Banjul,
Cairo, Dar es Salaam, Delhi, Freetown, Gaborone, Harare,
Hong Kong, Johannesburg, Kampala, Lagos, Lahore, Lusaka,
Mexico City, Nairobi, São Paulo, Tokyo*

Published in co-operation with the CTA (Technical
Centre for Agricultural and Rural Co-operation), P.O.B. 380,
6700 AJ Wageningen, The Netherlands.

ISBN 0–333–61202–7

Phototypeset by Intype, London

Printed in Hong Kong

A catalogue record for this book is available from
the British Library

The opinions expressed in this document and the
spellings of proper names and territorial boundaries
contained therein are solely the responsibility of the
author and in no way involve the official position
or the liability of the Technical Centre for
Agricultural and Rural Co-operation.

The ACP-EU Technical Centre for Agricultural and Rural Co-operation (CTA) operates under the Lomé Convention between Member States of the European Union and the African, Caribbean and Pacific States.

CTA collects, disseminates and facilitates the exchange of information on research, training and innovations in the spheres of agriculture and rural development and extension for the benefit of the ACP States.

To achieve this, CTA commissions and publishes studies; organises and supports conferences, workshops and seminars; publishes and co-publishes a wide range of books, proceedings, bibliographies and directories; strengthens documentation services in ACP countries; and offers an extensive information service.

Postal address: Postbus 380, 6700 AJ Wageningen, The Netherlands.

Agency for Cultural and Technical Co-operation (ACCT)

The Agency for Cultural and Technical Co-operation, an intergovernmental organisation set up by the Treaty of Niamey in March 1970, is an association of countries linked by their common usage of the French language, for the purposes of co-operation in the fields of education, culture, science and technology and, more generally, in all matters which contribute to the development of its Member States and to bringing peoples closer together.

The Agency's activities in the fields of scientific and technical co-operation for development are directed primarily towards the preparation, dissemination and exchange of scientific and technical information, drawing up an inventory of and exploiting natural resources, and the socio-economic advancement of young people and rural communities.

Member countries: Belgium, Benin, Burkina Faso, Burundi, Canada, Central African Republic, Chad, Comoros, Congo, Côte d'Ivoire, Djibouti, Dominica, France, Gabon, Guinea, Haiti, Lebanon, Luxembourg, Mali, Mauritius, Monaco, Niger, Rwanda, Senegal, Seychelles, Togo, Tunisia, Vanuatu, Vietnam, Zaire.

Associated States: Cameroon, Egypt, Guinea-Bissau, Laos, Mauritania, Morocco, St Lucia.

Participating governments: New Brunswick, Quebec.

Titles in *The Tropical Agriculturalist* series

Sheep	ISBN 0–333–52310–5	Ruminant Nutrition	0–333–57073–1
Pigs	0–333–52308–3	Animal Breeding	0–333–57298–X
Goats	0–333–52309–1	Animal Health Vol. 1	0–333–61202–7
Dairying	0–333–52313–X	Animal Health Vol. 2	0–333–57360–9
Poultry	0–333–52306–7	Warm-water	
Rabbits	0–333–52311–3	Crustaceans	0–333–57462–1
Draught Animals	0–333–52307–5	Livestock Production	
		Systems	0–333–60012–6

Upland Rice	0–333–44889–8	Sugar Cane	0–333–57075–8
Tea	0–333–54450–1	Maize	0–333–44404–3
Cotton	0–333–47280–2	Plantain Bananas	0–333–44813–8
Weed Control	0–333–54449–8	Coffee Growing	0–333–54451–X
Spice Plants	0–333–57460–5	Food Legumes	0–333–53850–1
Cocoa	0–333–57076–6	Cassava	0–333–47395–7
The Storage of Food		Sorghum	0–333–54452–8
Grains and Seeds	0–333–44827–8	Cut Flowers	0–333–62528–5
Avocado	0–333–57468–0		

Other titles published by Macmillan with CTA *(co-published in French by Maisonneuve et Larose)*

Animal Production in the Tropics and Subtropics	ISBN 0–333–53818–8
Coffee: The Plant and the Product	0–333–57296–3
The Tropical Vegetable Garden	0–333–57077–4
Controlling Crop Pests and Diseases	0–333–57216–5
Dryland Farming in Africa	0–333–47654–9
The Yam	0–333–57456–7

Land and Life series *(co-published with Terres et Vie)*

African Gardens and Orchards	0–333–49076–2
Vanishing Land and Water	0–333–44597–X
Ways of Water	0–333–57078–2
Agriculture in African Rural Communities	0–333–44595–3

Contents

Preface

This is the eleventh in the series of fifteen books on animal production in the tropics; (the ones already published are on poultry, sheep, pigs, rabbits, ruminant nutrition, dairy production, animal breeding, animal health (Volume 2), livestock production systems and goats). The series is intended to provide up-to-date information for students, extension specialists and farmers, written in an easy to understand manner. All the books are produced by specialists who have worked in a number of tropical countries or regions. This volume, the first of two, has been written by Archie Hunter, who is a senior lecturer at Edinburgh University's Centre for Tropical Veterinary Medicine. He has considerable experience of the problems caused by animal diseases in the tropics and has worked for long periods in several tropical countries.

Maintaining animal health in the tropics is a major problem – one to which most governments in tropical countries have given priority when attempting to improve the productivity of local stock. Because it is such a large subject it has needed two books in the series to cover all its aspects. This book, Volume 1, deals with the general principles of animal diseases and how they should be dealt with. Volume 2 covers individual diseases in detail.

This book explains the causes of animal diseases, how they are spread and the means available for their control. It covers not only diseases caused by micro-organisms, arthropods and helminths, but also deals with metabolic diseases and those caused by toxins. A unique feature is the inclusion of maps showing the geographical spread of all the important tropical diseases.

The signs of health and the symptoms to look for when an animal is suspected as being diseased are clearly presented. The book also has tables giving the main diagnostic and epidemiological features associated with individual diseases and metabolic conditions of the major farm animals found in the tropics, with the exception of poultry.

The book concludes by dealing with general veterinary procedures, including the administration of medicines. If it is read in conjunction with others in the series dealing with individual species and particularly with Animal Health Volume 2, readers will obtain invaluable information on the problems caused by animal disease in the tropics, their importance and control.

Anthony J. Smith

Acknowledgements

I am indebted to my collaborator, Professor Uilenberg, with whom it has been a pleasure to work, both for his thorough reading of the entire manuscript but also for his invaluable comments and advice. I am also grateful to my Edinburgh colleagues for their assistance, in particular Alan Walker, Duncan Brown, Martyn Edelsten, John Hammond and Chris Daborn. I am particularly indebted to Gordon Scott who has long shared with me the idea of developing simple diagnostic keys based on the main presenting clinical signs of disease; an idea which I hope has materialised to a certain extent in this book.

Photographs were kindly loaned by Richard Matthewman, Gavin Watkins, Chris Daborn, Peter Heath and Keith Sumption. Other photographs were either borrowed from the CTVM archives or are from my private collection.

Finally I would like to thank my wife Cynthia, both for spending long hours in front of our word processor, and for tolerating my anti-social behaviour while working on this volume.

Archie Hunter

1 Different types of diseases

Books on diseases of livestock often tend to concentrate on parasitic and infectious diseases. This is unfortunate because, as any experienced person appreciates, a high proportion of sick animals seen in the field do not have an infectious or parasitic disease but have something else wrong with them.

So what is a disease?
A disease is any process which disrupts an animal's normal function. Strictly speaking an animal with a broken leg has a disease because it cannot walk properly, but in practice we tend to exclude malfunctions caused by accidents when referring to diseases. There are several different types of disease and it is essential to appreciate these in order to understand specific aspects of individual diseases. Diseases can be classified in several ways and in this chapter the general principles of the different types of disease are outlined to correspond with the more detailed descriptions in Volume 2, namely infectious and contagious diseases, venereal and congenital infections, arthropods, arthropod-borne diseases, helminth infections and diseases associated with environmental and husbandry factors.

1 *Infectious and contagious diseases*

These terms are often used loosely for the same type of disease although they do mean different things. An infectious disease is one in which an animal is invaded by a foreign organism such as a virus, bacterium or parasite from another infected animal. Some infectious diseases require intermediary agents in order to spread from one animal; for example, equine infectious anaemia of horses can only be spread from one animal to another by blood-sucking biting flies. Other infectious diseases can spread between animals without any intermediary agents, and these are sometimes defined as contagious diseases, e.g. in contagious bovine pleuropneumonia (CBPP) animals are infected by inhalation of infected

1

droplets discharged by nearby clinical cases. Contagious diseases can be spread by direct transmission, such as CBPP mentioned above, or indirectly in which the infectious organism can survive outside the animal and be picked up from the environment; for example, the spores of the fungi that cause ringworm, a skin disease, can survive in the environment and be a source of infection to susceptible animals.

There are many different routes by which the organisms may pass from an infected animal to a non-infected one, and knowledge of these is essential to devising methods of preventing infection. The main routes of transmission are as follows.

1.1 Ingestion

In certain diseases, infected animals discharge the organisms to the environment. Susceptible animals may in turn become infected by ingestion of food or water contaminated with these discharges. This is an important route of transmission for many infectious organisms, and clearly the longer the organisms can survive in the environment, the greater the chances of susceptible animals becoming infected by this route.

Example – Foot-and-mouth disease One of the reasons why foot-and-mouth disease spreads so rapidly is because infected animals discharge large quantities of the infecting virus in saliva, milk, faeces, semen, urine and breath to the environment where it may survive for several months (Figs 1.1 and 1.2).

Fig 1.1   The saliva from this cow with foot-and-mouth disease contains large quantities of infectious virus.

Fig 1.2   Shared drinking troughs are one way in which animals can become infected by ingestion of water contaminated by clinical cases such as the animal in Fig. 1.1.

1.2   Inhalation

Animals may become infected by inhalation of infectious organisms that have been discharged into the air by infected animals.

Examples – Foot-and-mouth disease; contagious bovine pleuropneumonia   Because of the ability of foot-and-mouth virus to survive in the environment for prolonged periods, the virus can be carried by wind for long distances and susceptible animals many kilometres distant may become infected by inhalation of contaminated air.

Contagious bovine pleuropneumonia is an important respiratory disease of cattle and, as mentioned above, susceptible cattle become infected by inhalation of droplets containing the causative organism discharged by infected animals. Unlike foot-and-mouth disease virus, however, the causative organism survives only a few hours outside the animal and infection only occurs if there is close contact between sick and healthy animals.

1.3   Infection through the skin

Some organisms can infect animals via the skin, usually by contamination of cuts, abrasions, etc.

3

Example – Bovine farcy Bovine farcy is a chronic disease of cattle with lesions under the skin and infection probably occurs via contamination of skin wounds from tick bites, thorn bushes, etc. The causative organism may occur in the soil, although infected cattle are the most likely source of infection.

1.4 Infection from fomites

Fomites are any object which can convey infectious organisms, e.g. bedding, vehicles, harnesses, etc.

Example – Sheep and goat pox The virus that causes this serious skin disease can survive for many months in the environment. Animals can probably become infected by rubbing against contaminated fomites such as sheep pens etc.

2 *Venereal and congenital infectious diseases*

Venereal infections are transmitted by coitus (mating). In most venereal infections, the organism can be transmitted either way, i.e. from an infected male to a susceptible female or vice versa. Congenital infections are transferred from the mother to its offspring during pregnancy.

Example – Hog cholera In pregnant sows the virus that causes hog cholera can cross the placenta and infect foetuses which may be aborted or borne alive but deformed and trembling. This route of transmission is sometimes called vertical transmission.

3 *Arthropods*

In nature there are thousands of different species of arthropods including species of flies, ticks, lice, fleas, mites and bugs. Many are parasites of the skin of domestic animals; some are little more than a nuisance but others can cause serious skin irritation and damage. One important group of arthropods includes flies which deposit their eggs on animals (Fig 1.3). On hatching the fly larvae, or maggots, can penetrate skin, wounds and natural orifices. The resultant invasion of tissues is called myiasis and can be very distressing to the unfortunate animals attacked this way (Fig 1.4).

Some arthropod skin parasites are haematophagous, i.e. they have piercing mouth-parts with which they can pierce the skin and take blood meals for their nutrition. This feeding behaviour allows them to transmit a variety of infectious micro-organisms from animal to animal (see Chapter 2).

Most arthropods can be seen by the naked eye although detection of some mites, small arthropods that burrow into the skin and cause a disease called mange, requires microscopy. For more details of arthropods, readers are referred to Chapter 3 of Volume 2.

4 Arthropod-borne diseases

Many important infectious diseases are spread from animal to animal by biting arthropods. Thus an arthropod taking a blood meal from one infected animal may pick up the infection and transfer it to another animal at its next meal; thus the arthropods are the vehicles or vectors by which infection is spread from animal to animal. The infective agent commonly undergoes multiplication in the vector, which may then remain infective for long periods. However, transmission may also be 'mechanical' (see Chapter 2), when blood-sucking insects transfer infected blood directly from one animal to another, after their feeding on the first one has been interrupted.

Disease agents transmitted by arthropods include viruses, rickettsiae, protozoa and helminths. For example, small night flying blood-sucking species of *Culicoides* midges spread the virus that causes African horse sickness.

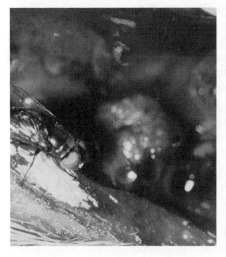 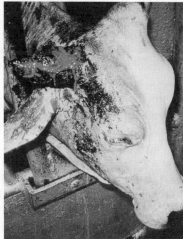

Fig 1.3. *Left* In the wounds of animals, *Chrysomya bezziana* (screw-worm fly) lays its eggs which hatch to larvae (CTVM).
Fig 1.4. *Right* Myiasis (horn strike) caused by screw-worm larvae (CTVM).

5 Helminth infections (worms)

Helminths (roundworms, thorny-headed worms and flat worms) of animals are major causes of disease and loss of productivity throughout the tropics. All types of domestic livestock are at risk to many different helminths which vary in their host range, life cycle and severity. Probably the most significant are those that parasitise the stomach and intestines but there are also important helminths of other parts of the body, e.g. lungs and liver.

6 Diseases associated with environmental and husbandry factors

Some important diseases of animals arise as a consequence of their environment or of the way they are managed by their owners. These include certain infectious diseases, metabolic disturbances, nutritional deficiencies and imbalances, poisons and a variety of miscellaneous conditions.

6.1 Infectious diseases

Certain infectious diseases are triggered by underlying environmental or husbandry factors. These are called predisposing factors and an understanding of them is very important when designing disease control programmes.

Example – Mastitis in modern dairy herds Mastitis is relatively rare in cows which suckle their calves naturally (Fig 1.5) but in high producing dairy cows milked by machine, it is a very serious problem. Unless done properly with high standards of hygiene, the process of machine milking can lead to bacterial infection of the udder and mastitis. In other words poor dairy husbandry is usually the predisposing factor of mastitis.

Example – Clostridial toxaemias One very important group of infectious diseases that are triggered by environmental or husbandry factors are clostridial toxaemias in which animals are poisoned by toxins produced by infecting clostridial bacteria. Clostridial bacteria are widespread in nature, being found in soil, organic matter and as natural inhabitants of the intestines of livestock. Infections with these organisms are normally harmless, but following certain predisposing factors, they can multiply rapidly and produce large quantities of toxins which poison the animal. The two main predisposing factors are a sudden improve-

Fig 1.5   This indigenous Yemen cow which suckles her calf is unlikely to suffer from mastitis or milk fever.

ment in nutrition, or tissue damage. Tetanus is a good example of the latter; the causative organism, *Clostridium tetani*, localises and multiplies in contaminated cuts and wounds producing a toxin which affects the nervous system.

6.2   Metabolic disturbances

These are often associated with intensive forms of animal husbandry and can be regarded as 'modern, man-made' diseases commonly associated with livestock industries in the USA, western Europe, etc. Nevertheless these diseases, sometimes called 'production diseases' can occur anywhere where intensive animal husbandry is practised and arise when there is an imbalance between dietary input and production offtake.

Example – Milk fever   In all cows around calving the onset of lactation causes a drop in blood calcium (hypocalcaemia) and in high producing dairy cows this can be excessive, resulting in milk fever which is frequently fatal unless treated quickly. Milk fever is virtually unknown in extensively grazed cattle which suckle their calves (Fig 1.5).

6.3   Nutritional deficiencies and imbalances

Ill health in animals is frequently due to deficiencies or imbalances in their nutrition. Sometimes the deficiency or imbalance will only cause disease if there is some other factor, for example, a phosphorous deficient diet may only cause ill health in cattle that are pregnant and/or lactating, but not in other cattle. In this case the combination of phosphorous deficiency plus pregnancy and/or lactation is required to cause disease.

It is often argued that nutritional deficiency (malnutrition) is the most important disease of livestock in the tropics. This may well be true but it is important to appreciate that animals suffering from malnutrition are more disease prone than those that are well nourished and hence more likely to be suffering from more than one disease simultaneously. This can cause difficulties in diagnosis; thus an animal in poor condition may be found to be infected with large numbers of internal parasites and treated accordingly but when that same animal fails to recover satisfactorily, only then may it be appreciated that the internal parasite infection was secondary to the main underlying problem of malnutrition.

6.4 Poisons and toxins

Diseases caused by poisoning are amongst the most troublesome to diagnose because of the technical difficulties of detecting and identifying poisons. There are very broadly two types of poisons: those that are produced biologically are referred to as toxins, e.g. by plants, microorganisms, etc., and poisons, which by definition refer to non-biological sources such as chemicals, etc. In reality the two terms, poisons and toxins, are often used loosely as though they mean the same thing. The main sources of poisons and toxins to domestic animals are plants, certain feeds, mouldy feed, and farm chemicals.

Plants There are hundreds of species of plants found in pastures that are poisonous to livestock. Fortunately they are normally avoided by livestock, but under certain conditions they may be eaten, such as during a drought when grazing is scarce. Plant poisoning is an important cause of illness and death in tropical countries, and unfortunately the plants concerned are often poorly documented. A few of the known important poisonous plants are briefly outlined in Volume 2, but a comprehensive account would be a book in itself.

Insecticides and acaricides An important potential source of poisons to livestock are the chemicals used for treating skin arthropod parasites, e.g. dip chemicals for ticks. These are well documented and some of the commonly used ones are described briefly in Volume 2.

Botulism One important poisoning is botulism which can be a serious cause of illness and death of livestock in the tropics. Botulism results from the ingestion of food or water contaminated with a toxin produced by the bacterium *Clostridium botulinum.*

7 Inherited traits

7.1 Congenital diseases

Not included in Volume 2, and not to be confused with congenital infections, are diseases inherited as a direct result of an animal's genetic make-up or genotype. Very loosely, an animal's genotype can be described as comprising pairs of genes located in the chromosomes, one of the pair being inherited from the father and the other from the mother. For more details readers are referred to *Animal Breeding* by G. Wiener in this series. The continuation of genes which produce harmful or disease-associated traits is largely curtailed by evolutionary processes, except those which are dominated by their normal opposite pairing gene. These genes are called recessive genes and their effects only appear when they are paired with the same gene. Unfortunately, the effects of some pairs of recessive genes are physical or metabolic defects (Fig 1.6). Clearly both the parents have to carry the gene in question although they themselves may be normal. Fortunately such inherited recessive gene-associated disorders are rare, for example, goitre, or enlargement of the thyroid gland, may be caused in certain breeds of cattle and small ruminants by inheritance of recessive genes.

NB There are other causes of goitre.

Fig 1.6   Calf with hydrocephalus, a rare congenital disease, in which the calf was born with a defect in the circulation of the central nervous system resulting in accumulation of fluid in the brain.

7.2 Inherited resistance to disease

Of much greater importance than inherited physical or metabolic disorders is the inherited ability of certain breeds of animals to resist disease better than other breeds. A good example is the recognised ability of certain dwarf breeds of cattle in West Africa (e.g. N'Dama, Bauolé and Muturu, Fig 1.7) to resist tsetse fly transmitted trypanosomosis, whereas virtually all other cattle breeds are susceptible (Fig 1.8). Throughout much of the tropics, local cattle populations are well known for their ability to resist ticks and tick-borne diseases better than, say, European breeds and this trait is now being exploited in cattle breeding programmes.

Fig 1.7   Dwarf West African Muturu cow which is resistant to tsetse fly transmitted trypanosomosis (Richard Matthewman).

Fig 1.8   Indigenous Botswana cattle in poor condition because they are suffering from tsetse fly transmitted trypanosomosis.

8 *Disease definitions*

When describing diseases and how they develop, there are a number of terms that are used routinely. Some of the more common ones are explained below as they are referred to in Volume 2.

8.1 Acute and chronic diseases

Most diseases have a pattern of development of the sequence of events that make up the disease picture. Thus a disease in which the sequence develops rapidly is described as acute, whereas a chronic disease is one in which the sequence is drawn out over a prolonged period. It is often useful to know whether a disease is acute or chronic because this may be a valuable diagnostic aid. Acute diseases may be very rapid (peracute) or slower (subacute).

Examples – Anthrax, rinderpest and trypanosomosis Anthrax is a very rapid, usually fatal, bacterial infection of livestock. It may be so rapid that infected animals are found dead without prior observations of any clinical signs. This form of anthrax would be described as peracute.

Rinderpest is a highly fatal contagious viral disease of livestock, particularly cattle. Although the disease can follow a peracute course with affected animals dying within a few days of the onset of symptoms, it is more typically seen in the acute form in which symptoms last for approximately one to two weeks before death. Trypanosomosis, however, an important blood parasitic disease of livestock throughout the tropics, is usually a chronic disease with animals suffering for months or even years.

8.2 Sub-clinical infections

A difficult concept for many people is the known ability of animals to harbour potentially harmful organisms without suffering any overt signs of disease. For example grazing livestock worldwide harbour helminths in their stomachs and intestines and only become sick when these burdens reach significant levels. The presence of low sub-clinical worm burdens is thus quite natural and nothing to worry about, provided animals are not subjected to conditions that allow these levels to build up to significant levels.

Other good examples are certain tick-borne blood infections such as anaplasmosis, babesiosis and theileriosis. Throughout much of the tropics and sub-tropics, indigenous livestock may be infected with these from an early age and remain infected throughout life without suffering any ill effects. An awareness of such widespread sub-clinical infections is very important because introduced exotic animals are usually susceptible to these infections and, if attacked by the same ticks that attack local animals, are liable to suffer severe disease as a result.

9 Recognition of different types of disease

A great deal of skill and training is involved in the identification of different types of disease. A veterinary undergraduate spends approximately five years at university acquiring the necessary skills and knowledge for this, so it is unrealistic to expect lay staff, farmers and other personnel working with livestock to be able to diagnose livestock diseases other than those which are common and well known. Nevertheless, much of disease diagnosis depends on simple common sense. Indeed in certain situations it may be necessary for lay personnel to attempt a diagnosis in order to take immediate action pending the visit of a professionally trained veterinarian. In such cases the first step would be to identify which type of disease is causing the outbreak and the following can be used as guidelines:

Infectious and contagious diseases; arthropod borne diseases; helminth infections
• More than one animal affected.
• Animals of different age groups and sex may be affected.
• Arthropod-borne diseases and helminth infections are often seasonal.

Venereal infections
• Confined to breeding adults.

Congenital infections
• Disease effects usually apparent at birth or shortly after.

Congenital disease
• Metabolic or physical defect at birth. Usually rare.

Diseases associated with environmental and husbandry factors
• Usually confined to animals associated with a specific type of husbandry or production.

Poisons
• Accidental poisonings are usually single episodes.
• Plant poisonings often preceded by change in grazing/browsing pattern; the change may be enforcement of animals to eat plants normally avoided because of lack of conventional fodder, e.g. in drought, overgrazing, etc.

NB The above are only guidelines and must not be regarded as definitive steps to making a diagnosis. In nature there are many exceptions to general rules and diseases are no different, being accidents of nature!

2 Arthropods and helminths

Arthropods and helminths that infest livestock in some way are usually defined as parasites. Strictly speaking, a parasite is an organism that lives in or on another organism, and so many micro-organisms (see Chapter 3) could be defined as parasites. In practice, however, only arthropods, helminths and protozoa (the most complex micro-organisms) are defined as parasites.

1 Arthropods

As the name implies, arthropods are organisms with jointed legs and those of veterinary importance belong to two groups, insects and acarines.

1.1 Insects (flies, lice and fleas)

Insects typically have a head, thorax supporting three pairs of legs, and abdomen. Many are winged and can fly.

Flies of veterinary importance can all be seen by the naked eye but range greatly in size. At one end of the size range are black flies (*Simulium* spp.), small flies of 1.5 to 5 mm long which can attack livestock in swarms and cause painful bites and blood loss. By contrast tabanids, or horseflies as they are often called, are the 'Jumbo jets' of the fly world reaching up to 25 mm long. These large flies can cause deep painful bites and their blood-sucking habits make them effective vectors of a number of important pathogenic organisms. Examples of flies of veterinary importance are illustrated in Fig 3.1 (Volume 2).

Lice, unlike flies, are wingless and are true parasites in that they live entirely on the skin of their hosts, being unable to survive more than a day or so off them. They are flattened from top to bottom and about 1 to 5 mm long, being much smaller than flies (see Fig 3.4, Volume 2). Biting or chewing lice live off scales of skin, scabs, and the outer surface of hair, wool and feathers. Sucking lice, unlike biting lice, are confined

13

to mammals and have mouthparts which can pierce skin and suck blood. Under certain conditions, lice burdens can build up to significant levels and cause considerable irritation and, in the case of sucking lice, blood loss. The scientific term for this is pediculosis.

Fleas are also wingless but, unlike lice, spend most of their life off their hosts. They are flattened from side to side and have powerful hind legs which make them the champion 'high jumpers' of the insect world. Whenever they need a blood meal, they literally hop on to their hosts, causing irritation in the process. Under certain conditions, they commonly attack humans, cats, dogs, pigs and poultry, but ruminants and equids much less frequently.

1.2 Acarines (ticks and mites)

These are wingless and do not have such clearly defined body components as insects. Their bodies comprise a front section with mouth parts, and a main body which supports the legs, three pairs in larvae and four pairs in nymphs and adults. Of the two kinds of ticks, hard and soft, the former is the more important.

Hard ticks   These are oval shaped, flattened dorso-ventrally, and have a hard protective shell on their dorsal surface called a scutum. After taking a blood meal, adult females detach from their host and lay large numbers of eggs which hatch to small larvae, which have three pairs of legs. These can live unfed for several months in the environment, during which period they have to attach to a host and take a blood meal in order to survive. Following their blood meal, larvae moult to nymphs which resemble small adults. Nymphs in turn must find a host and take a blood meal before moulting to adults. The cycle of attaching to hosts and taking blood meals means that hard ticks are very important vectors of several major blood-borne diseases of livestock.

In addition, ticks themselves are disease agents. Some species of ticks have particularly long mouthparts and can cause considerable skin damage. If ticks are present in large numbers, the constant loss of blood can affect the health of the host. Examples of hard ticks are shown in Figs 3.5 to 3.8 (Volume 2).

Because hard ticks have to spend considerable periods in the environment waiting for a passing host to attach to, they tend to be found in habitats that protect them from climatic extremes. This is usually bushy or grass cover, but some species of hard ticks can also survive in cracks in walls and buildings. A knowledge of their habitat requirements is important when devising tick control programmes. More details of this aspect are provided in Volume 2.

Soft ticks Unlike hard ticks, soft ticks do not possess a scutum. Their blood meals tend to be moderate and frequent, and so soft ticks are often found near resting places of livestock, for example, under shade trees, in paddocks and huts, etc.

Mites These are the smallest of the parasitic arthropods, most being less than 0.3 mm long and barely visible to the naked eye. Like lice, many are true parasites and spend all their life on the skin of their hosts. They thus spread from animal to animal by contact, and once infected, a population of mites can build up to pathological levels on the same host without further infection from another host. The skin irritation and damage caused by these parasitic mites is called mange, and this varies in severity in different species (Fig 2.1).

Not all mites are parasitic, however, and slow moving oribatid mites found commonly in pastures worldwide, can transmit tapeworms to livestock. Tapeworm stages passed to the environment in the faeces of infected livestock may be ingested by the mites in which they develop to larvae. If the mites are in turn ingested by livestock, the larvae are released in the intestines and complete their development in the animal host.

Fig 2.1 Sarcoptic mange in goats caused by *Sarcoptes scabiei* (Gavin Watkins).

2 Transmission of diseases by arthropods

Of the arthropods that parasitise the skin of livestock, flies and ticks are the most important disease vectors, and arthropod-borne diseases in Volume 2 are grouped into those that are transmitted by flies, and those that are transmitted by ticks.

There are two ways in which disease agents can be transmitted by arthropod vectors, mechanical transmission and cyclical transmission.

2.1 Mechanical transmission

This occurs when the arthropod vector simply transfers the disease agent from one animal to another, usually on its mouthparts, with no development of the agent in the arthropod, e.g. *Trypanosoma evansi*, an important blood parasite of domestic animals throughout much of the tropics, which is transmitted mechanically from animal to animal by blood-sucking flies. There are several other important diseases that are spread mechanically by blood-sucking arthropods and it is important to appreciate that as the arthropods are simply acting as vehicles of infected blood, this can be done by other means, for example, by blood-contaminated syringe needles, castration knives, etc.

2.2 Cyclical transmission

This occurs when disease agents complete part of their development within the arthropod vectors that transmit them, such as tsetse fly transmitted trypanosomosis of livestock in sub-Saharan Africa. When tsetse flies take a blood meal from infected animals, the trypanosome blood parasites continue multiplying and develop to infectious forms within the flies, which can then infect other animals at their subsequent blood meals.

In this form of transmission, the vectors are an integral part of the life cycle of the disease agent, hence the term cyclical transmission. Although disease agents that are transmitted cyclically by arthropod vectors can sometimes also be transmitted mechanically by other vectors, in practice this is usually of little importance.

3 Helminths (worms)

As mentioned in the previous chapter, there are three kinds of worms that parasitise livestock: roundworms, thornyheaded worms and flatworms.

3.1 Roundworms (nematodes)

As the name implies, these worms are round in cross section, giving them a cylindrical shape tapering at both ends. They have a simple tubular digestive system with a buccal cavity (mouth) and anus (females) or cloaca (males). They vary greatly in size; at one end of the spectrum is *Ascaris suum*, a 40 cm long worm of the small intestine of pigs; at the other end are species of *Trichostrongylus*, parasites of the stomach and intestines of various livestock, the adults of which are less than 7 mm long and barely visible to the naked eye.

Although different species of roundworms parasitise various parts of the body of domestic livestock, all have the same basic life cycle. The sexes are separate and females lay large numbers of eggs from which hatch small larvae which have the same basic shape as the adult roundworm. The immature larvae moult four times, and it is conventional to refer to the five larval stages as L_1, L_2, L_3, L_4 and L_5, the last of which is an immature adult. Although there are many variations, in the life cycle of all roundworms eggs or hatched larvae are passed to the environment from an infected animal, usually in the faeces. If conditions are suitable, there is further development to the infectious form of the roundworm, which is then available in the environment to infect other animals. Variations in the life cycle are outlined below.

Roundworms in which L_3 are infectious Many of the most important roundworms of the stomach and intestines of domestic livestock have the following life cycle. Eggs of adult female roundworms in the stomach or intestines are passed in the faeces to the environment, where they hatch to L_1 which develop to infectious L_3 following two moults. If ingested by a susceptible host, the L_3 develop to adults in the stomach or intestines following another two moults. A great deal of research has gone into the various stages of the above cycle, which is illustrated diagrammatically in Fig 5.1, Volume 2.

A key component is the development outside the host of eggs to L_3 and ability of L_3 to survive in the environment long enough to have a reasonable chance of being picked up by a susceptible host. If it is too hot (over 26°C) larvae will hatch and moult quickly but are unlikely to develop to L_3; if it is too cold (below 10°C), hatching of eggs is arrested. Humidity is also important and the ideal conditions for development and survival of infectious L_3 larvae are pastures in warm humid climates (Fig 2.2).

Variations in roundworms with infectious L_3 – lungworms, *Strongyloides* worms and the pig kidney worm In the life cycle of lungworms of ruminants, L_1 larvae, not eggs, are passed in the faeces of the host. Adult

17

Fig 2.2   Small ruminants grazing in the humid tropics, ideal conditions for stomach and intestinal roundworms (Chris Daborn).

female worms in the airways of the host produce eggs which hatch immediately releasing L_1 larvae which are coughed up, swallowed and passed out to the environment in the faeces where they develop to infectious L_3 in the normal way. If ingested, L_3 larvae moult and migrate from the intestines via liver and lungs to airways of the lungs where they complete their development to adult worms.

Another variation to the basic life cycle is that of *Strongyloides* species, parasites of the intestines of domestic animals. Eggs passed in the host's faeces hatch and produce L_3 larvae in the normal way. These can either infect hosts or continue their development as free-living nematodes in the environment. As well as infecting hosts following ingestion, L_3 larvae on pasture can penetrate skin and migrate to the small intestine via the circulation, lungs and trachea.

Yet another variation is the unusual life cycle of *Stephanurus dentatus*, the pig kidney worm. In this case, eggs are passed in the host's urine to the environment where they can develop to infectious L_3 in the normal way or in earthworms which ingest the eggs. Infection can follow ingestion of L_3, either direct from the environment or in earthworms, or by L_3 penetration of the skin (cf. *Strongyloides*). Following infection, larvae moult and migrate to tissues surrounding the kidneys, either via the liver following ingestion, or via the bloodstream, lungs and liver following skin penetration. The fully developed adult worms become enclosed in a capsule and their eggs are passed into the ureter, the tube that passes urine from the kidneys to the bladder.

18

Ascarid worms   These large white roundworms of domestic animals are parasites of the small intestines (Fig 2.3). Infection can result from several ways, the commonest being by ingestion of the thick-shelled eggs containing L_2 passed in the faeces. The eggshell is very resistant to temperature extremes and eggs can remain viable in the environment for several years, protecting the infectious L_2 inside. Following ingestion, the eggs hatch in the small intestine, releasing the L_2 which moult and migrate via the liver, bloodstream, lungs, trachea and back to the small intestine where they complete their development to adults. Migrating larvae may cause some tissue damage resulting in scar formation, such as 'milk spots' in the liver (Fig 2.4). The eggs of some of these worms may be ingested by transport hosts in which they hatch to infectious L_2 larvae. For example, earthworms or dung beetles may ingest the eggs of *Ascaris suum*, and pigs may become infected by ingesting these in turn.

Infection with these worms is common, but in adults they are of little or no clinical significance and larvae may remain dormant in tissues. In pregnancy, however, larvae become active and can infect the foetus or migrate to the mammary gland and be passed in milk to the sucking newborn. Buffalo calves can be infected with *Toxocara vitulorum* by both of these routes, the latter being particularly important.

3.2   Thornyheaded worms

There is only one species of veterinary importance, *Macracanthorhynchus hirudinaceus*, a parasite of the small intestine of pigs. These are very large worms, the females reaching up to 65 cm long. Eggs passed in the faeces are very resistant to climatic extremes and can survive for several years in the environment. If ingested by dung beetle larvae, they develop to the infectious stage and pigs become infected by ingestion of beetle grubs or adults.

Fig 2.3. *Left*   Large ascarid roundworms in the intestine of a calf (CTVM).
Fig 2.4. *Right*   'Milk spot' scars in pig liver caused by earlier migration of ascarid larvae (CTVM).

19

3.3 Flatworms

These include tapeworms and flukes.

Tapeworms Adult tapeworms are parasites of the intestines of their hosts. They have a head, called a scolex, which attaches by suckers and sometimes hooks to the lining of the intestines, although one species, *Stilesia hepatica* is found in the bile ducts of ruminants. Segments bud from the 'neck' of the scolex so that as the tapeworm develops, it becomes longer with a tape-like body comprising a chain of segments, the mature segments at the tail of the worm being larger than the newly developed ones at the scolex. Each segment contains both male and female organs (hermaphrodites), and as they mature the uterus becomes filled with eggs. The fully mature, or gravid, segments break off and are passed in the faeces to the environment where they release the eggs.

To develop further, the eggs must be ingested by another host (intermediate host). Within the intermediate host, eggs develop to larval stages which consist essentially of immature scolices contained in a cyst in the tissues of the intermediate host. Following ingestion of these tissues by the host, the cyst wall breaks down and releases the immature scolex which attaches to the intestine wall. The hosts and intermediate hosts of tapeworms are shown in Table 2.1.

As can be seen from Table 2.1, adult tapeworms vary greatly in length. The final hosts and intermediate hosts of tapeworms can be classified as two types; in one type the final hosts are herbivorous animals who become infected by ingesting the intermediate hosts, forage mites, from pastures or vegetation; in the other type the final hosts are carnivorous

Fig 2.5 *Taenia saginata* cysts in bovine heart muscle. If ingested by humans, these cysts develop to tapeworms in the intestines (CTVM).

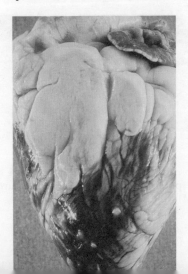

Table 2.1 Tapeworms of domestic animals and man

Adult tapeworm	Length	Final host	Intermediate host
Anoplocephala perfoliata	20 cm	Equines	Forage mites
Anoplocephala magna	80 cm	Equines	Forage mites
Paranoplocephala mamillana	5 cm	Equines	Forage mites
Moniezia benedeni	2 m+	Cattle	Forage mites
Moniezia expansa	2 m+	Ruminants and camels	Forage mites
Stilesia hepatica	50 cm	Ruminants	Forage mites?
Avitellina spp.	3 m	Ruminants and camels	Forage mites and lice?
Echinococcus granulosus	6 mm	Dogs	Ruminants, camels and man
Taenia multiceps	1 m	Dogs	Ruminants
Taenia saginata	5 m+	Man	Cattle
Taenia solium	5 m+	Man	Pigs

and become infected by ingestion of cyst-infected tissues of the intermediate host (Figs 2.5 and 2.6). Knowledge of these hosts is essential in developing control measures as described in Chapter 5 of Volume 2.

Flukes These worms are flattened dorso-ventrally. Those of veterinary importance are parasites of liver bile ducts, alimentary systems and blood vessels. Like tapeworms, many are hermaphrodite and their eggs are passed in either faeces or urine to complete their development in the intermediate hosts which are various kinds of snails (Fig 5.3, Volume 2).

As with tapeworms, an understanding of these life cycles is essential to

Fig 2.6 Butchering goats at a village slaughter point. Dogs in the background have access to viscera which are liable to contain cysts of the tapeworms *Echinococcus granulosus* and/or *Taenia multiceps*.

Fig 2.7 Sheep grazing and drinking at an irrigation channel heavily infested with snails (*Lymnaea truncatula*), the intermediate host of *Fasciola hepatica*. Liver fluke was controlled in this flock by changing to piped irrigation (Peter Heath).

devising control programmes. For instance, *Fasciola hepatica* causes liver fluke, a very important debilitating disease of domestic animals. If the habitat of the intermediate host 'aquatic snails' can be identified, it may be possible to prevent susceptible animals from grazing there (Fig 2.7).

4   How do helminths cause disease?

Although there are a number of different types of helminths and helminth life cycles, they have relatively few ways of causing disease, as indicated below.

4.1   Loss of condition due to competition for nutrients

Many of the helminths of veterinary importance are parasites of the stomach and intestines. They derive their nutrients from the contents of the alimentary system, and if present in large numbers, can significantly deprive the host of the same nutrients, leading to loss of condition. This is possibly the most important pathogenic effect of roundworms and, to a much less degree, tapeworms.

4.2 Diarrhoea due to gastro-enteritis

Helminths of the stomach and intestines of domestic animals can cause varying degrees of inflammation to the lining of the stomach and intestines, resulting in diarrhoea leading to an additional loss of nutrients which pass too quickly to be digested. Roundworms are particularly important in this respect (Fig 2.8).

4.3 Anaemia due to blood loss

Some helminths suck blood and if present in large enough numbers, can cause significant blood loss to the host and anaemia. Particularly important in this respect are *Haemonchus* species, very important roundworms of the stomach of ruminants and camels.

4.4 Tissue damage from migrating larvae

As can be seen from above, the larvae of many helminths migrate through the tissues of their hosts before reaching their final destination to complete their development to adults, causing varying degrees of tissue damage en route. The pig kidney worm, *Stephanurus dentatus*, is an extreme example of this; heavy infections can cause severe liver damage and even liver failure and death.

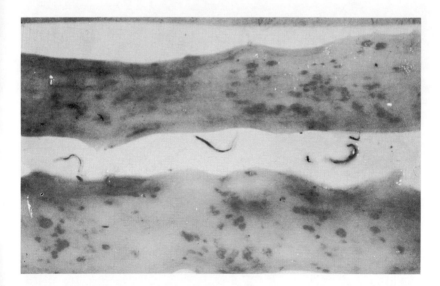

Fig 2.8 Haemorrhages and ulceration in the lining of the intestine of sheep, caused by infection with the roundworm *Bunostomum trigonocephalum* (CTVM).

23

It must be appreciated, however, that tissue damage from migrating larvae often results in little or no clinical signs. The damage is often only detected as minimal scarring of tissues at slaughter in the abattoir or slaughter point, such as the example of 'milk spot' liver in Fig 2.4.

4.5 Reaction to helminths in tissues

As well as parasitising the stomach or intestines, some helminths localise in various tissue sites, provoking various degrees of reaction. At one end of the spectrum are *Onchocerca* worms which are 'injected' into the skin of domestic animals by biting flies and cause a very mild skin reaction; see 'worm nodule disease' in Volume 2. By contrast, severe reaction to *Dictyocaulus* worms in the airways of domestic animals can cause bronchitis which may lead to pneumonia by secondary bacterial infections.

3 Infectious diseases

In order to understand better the processes by which infectious diseases occur, it is essential to appreciate the actual agents that cause them. We have already seen that infectious diseases can spread from animal to animal by various means including transmission by arthropod vectors, and that infectious diseases that spread between animals, either directly or indirectly, are defined as contagious. Irrespective of the means of transmission, the actual agents that cause these diseases are various types of micro-organisms.

Micro-organisms, as the name implies, are very small; so small in fact that they can only be seen with the aid of a microscope. Because these micro-organisms cause a range of pathological changes to tissues once they have invaded an animal, they are defined as being pathogenic and are often referred to as pathogens.

It is essential to appreciate that of the thousands of different types of micro-organisms in nature, only a small proportion are pathogens, and that indeed some are essential to the health of animals and people. For example, the digestion of our food depends on the presence in the stomach and intestines of micro-organisms which assist in breaking down the food to elements which can be absorbed into our body systems, particularly in ruminants.

There are four types of pathogenic micro-organisms: viruses, bacteria, fungi and protozoa.

1 Viruses

These are the smallest of the pathogens, so small that they can only be seen by very powerful electron-microscopes. Essentially they consist of genetic material (nucleic acid) protected by a protein coat (capsid). Some viruses have an outer lipid envelope around the capsid. Because they are very basic organisms, they only replicate by invading cells of higher organisms and 'high-jacking' their metabolic processes. As a consequence, cells are 'instructed' by the invading virus to manufacture more virus particles. The invaded host-cell may or may not be destroyed when the progeny viruses are released to invade more cells. Some viruses can do this on a large scale and the degree of cellular damage that viruses can cause is referred to as their virulence.

Thus viruses are obligatory intracellular organisms and invade cells of all forms of life, including animals, insects, plants, bacteria and fungi. Those pathogenic viruses of veterinary interest invade the cells of animals and damage them in the process. For example, rinderpest virus invades cells of the alimentary and respiratory tracts and the resultant cell damage causes diarrhoea and discharges from the mouth and nose (Fig 1.16, Volume 2). An additional important feature of rinderpest is that the virus also destroys the cells that are responsible for the host's defence mechanisms, making the host more vulnerable to other less pathogenic organisms. Another example is foot-and-mouth disease virus which invades various tissues including the surface of the tongue (Fig 3.1).

It has been estimated that viruses account for up to 60 per cent of outbreaks of disease in animals and man. Despite their major importance, very few anti-viral drugs have been developed. The problem lies in the close association between viruses and host cells and that any drugs lethal to viruses are liable to be harmful to host cells as well. Their cost generally limits their use to man but they are sometimes used to treat animals, e.g. for orf.

Fig 3.1   Foot-and-mouth disease. The infecting virus has damaged the surface of the tongue resulting in multiple-ruptured vesicles.

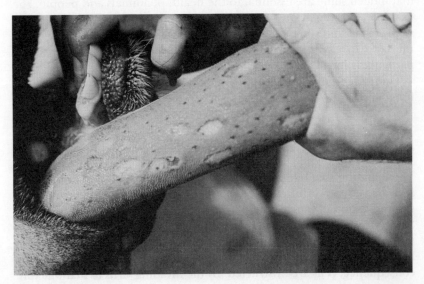

2   Bacteria

Bacteria are single-celled organisms of a higher order than viruses. Like viruses, some bacteria are obligate intracellular organisms, while others thrive quite happily extracellularly in the environment. Although readers of this book may not be interested in the classification of different bacteria, a little explanation is helpful in understanding how they behave and cause disease. There are four types of bacteria in approximate ascending order of size: chlamydia, rickettsia, mycoplasma and true bacteria.

2.1   Chlamydia

Like the true bacteria, they have cell walls but do not possess all the necessary metabolic processes to replicate themselves and consequently, like viruses, are obligate intracellular parasites. They are round shaped and have a predilection for epithelial cells, i.e. cells of the lining of mucous membranes (see Chapter 4). *Chlamydia psittaci* is an example, which causes enzootic abortion of sheep and goats, and infects the placenta of pregnant ewes causing placentitis and abortion.

2.2   Rickettsia

These are very similar to chlamydia and are obligate intracellular microorganisms. Some rickettsia invade the cells lining blood vessels (endothelial cells) where they multiply and break out into the bloodstream and spread around the body. Blood vessels are liable to be blocked or damaged as a result. *Cowdria ruminantium* which causes heartwater is a good example of this type of infection. The damage to blood vessels caused by the *Cowdria* organisms causes leakage into various body cavities including the sac surrounding the heart (Fig 3.2). One rickettsial organism, *Anaplasma*, invades and destroys red blood cells causing anaemia.

Although not a feature of all rickettsial infections, many of those of importance in livestock in the tropics are transmitted by ticks, e.g. anaplasmosis, heartwater, ehrlichiosis and cytoecetosis.

2.3   Mycoplasma

These are the smallest and simplest micro-organisms that can replicate by themselves without recourse to invading cells of higher organisms. Unlike other bacteria, they do not have a rigid cell wall but are surrounded by a flexible membrane which allows them to change their shape.

Fig 3.2 Heartwater. The sac protecting the heart has been opened to reveal excess surrounding fluid (Keith Sumption).

Unlike viruses, they are extracellular and harmless mycoplasmas are found in the digestive, respiratory and genital tracts of animals. Pathogenic mycoplasmas tend to have a predilection for certain tissues. *Mycoplasma mycoides* subsp. *mycoides*, the cause of contagious bovine pleuropneumonia, localises in the chest causing inflammation of the lungs and their lining (pleura) (see Fig 1.2, Volume 2). Mycoplasma infections are often low grade and chronic.

2.4 True bacteria

True bacteria are the largest of this group of micro-organisms. Being the highest order of bacteria, they possess all the metabolic processes necessary for replication and so do not have to invade cells. They have rigid cell walls which give each species its characteristic size and shape. The three basic forms of bacteria are rod-shaped (bacillus), round (coccus) and spiral or curved rods (spirochete and vibrio), although there is a range of intermediate forms, e.g. roundish rod shaped bacteria are defined as coccobacilli. Because their requirements for replication are not so specific as viruses, chlamydia and rickettsia, bacteria are relatively easy to grow on artificial laboratory medium and pose fewer diagnostic problems when it comes to isolating them from clinical cases of disease.

Bacteria cause disease by their ability to invade tissues, and their production of toxins, properties which vary widely from species to species and determine the degree of their virulence.

Invasive bacteria do so by enzymes which break down host tissues and allow the bacteria to penetrate and spread. An extreme example of bacterial invasiveness is the frequently fatal disease anthrax, caused by

Fig 3.3   Anthrax in an ox. The anthrax bacillus is very invasive and spreads throughout the body causing haemorrhages from the orifices including the nostrils as shown (CTVM).

Bacillus anthracis, which penetrates all body tissues prior to death (Fig 3.3).

Not all pathogenic bacteria are invasive, however, and others with little or no invasive ability produce their harmful effects by secretion of poisons called exotoxins. The poisoning ability, or toxicity, of these bacteria varies widely and one species produces one of the most lethal poisons known to man, namely *Clostridium botulinum.* Fortunately botulism, the name of the disease caused by poisoning with *C. botulinum* toxin, is not very common and where it is a problem, animals at risk can be vaccinated. Another example is diarrhoea in young livestock caused by toxin producing strains of *Escherichia coli.* These bacteria often infect the intestines of livestock and under certain circumstances, multiply rapidly producing large quantities of exotoxins which can be fatal (Fig 3.4).

Fig 3.4   Acute enteritis in a lamb caused by infection of the intestines with a toxin producing strain of *Escherichia coli.*

3 Fungi

Fungi are widespread in nature and well known to all of us. Moulds that form on stale food are fungi, as are the many different types of toadstools and mushrooms. Their structure is similar to plants in some respects, but unlike plants they are not photosynthetic, i.e. they are unable to utilise the energy of sunlight. Hence they are found growing on material from which they can derive their nutrients such as organic matter, vegetation and sometimes animals.

Fungi can be classified into two basic types, moulds and yeasts. Moulds grow as colonies of many celled filaments, whereas yeasts grow as individual round or oval shaped cells. Some fungi can grow as either yeasts or moulds depending on conditions.

3.1 Pathogenic fungi

A few species of fungi are pathogenic to animals. For example, ringworm is a skin infection of all domestic animals caused by species of *Trichophyton* or *Microsporum*, mould type fungi. Similarly epizootic lymphangitis is a skin infection of horses caused by *Histoplasma farciminosum* (see Fig 1.9, Volume 2).

3.2 Opportunistic pathogens

Some fungi can be opportunistic pathogens, e.g. following prolonged treatment with antibiotics (see Chapter 7), the natural bacterial population of an animal's intestines may be depleted and create conditions favourable for colonisation by normally harmless fungi. This is one of the reasons why veterinarians and doctors endeavour to limit the use of antibiotics when treating bacterial diseases.

3.3 Mycotoxicoses

Certain cases of poisoning are caused by toxins produced by fungi, called mycotoxins. The resultant diseases, called mycotoxicoses, usually result from eating stale or wet food that has become mouldy and can thus often be attributed to poor husbandry. Ergotism, a disease in which the tissues of the extremities (ears, lips, etc.) are destroyed and which can affect all domestic livestock, is caused by a toxin produced by *Pithomyces chartarum* which can grow on cereals, usually in warm wet conditions (see Chapter 6, Volume 2).

4 Protozoa

This is arguably the most complex group of infectious micro-organisms. Like other micro-organisms, many thousands of species of protozoa are found in nature, of which a small proportion infect animals and only a few of these are pathogenic. Protozoa are single-celled organisms and pathogenic protozoa of veterinary interest belong to two main groups, the flagellates and the apicomplexans.

4.1 Flagellates

These possess whip-like structures called flagellae which the protozoa can use to propel themselves through fluids. Hence these organisms are found outside cells, and have leaf-shaped bodies suitable for swimming around in body fluids such as blood plasma, e.g. *Trichomonas foetus* and various species of *Trypanosoma*.

Trypanosomoses are very important fly-transmitted flagellates of all domestic livestock in which the trypanosomes multiply in blood plasma and sometimes in fluids of other tissues such as the brain or the eye. Trichomonosis is a venereal infection of cattle caused by *Trichomonas foetus* in which the protozoal organisms are found in the fluids of the genitalia.

4.2 Apicomplexans

These are very complex intracellular organisms that have complicated life cycles with sexual and asexual forms of multiplication. Some apicomplexans invade and multiply inside animal gut cells, damaging them in the process, eventually producing infectious eggs (oocysts) which are passed out in the faeces to the environment where they can be picked up by other susceptible animals. Examples of infectious pathogenic apicomplexans that are spread from animal to animal this way are the protozoa that cause coccidiosis and cryptosporidiosis, both enteric diseases of young animals.

One very important group of pathogenic apicomplexans, *Theileria* and *Babesia* species, do not develop oocysts but are spread to animals by cyclical transmission in ticks (see Chapter 2). Following the bite of an infected tick, the organisms eventually invade cells in the bloodstream of the animal at which point the animal is infectious to ticks. In any ticks feeding on such an animal, the organisms then develop to infectious forms in the salivary glands so completing the life cycle. It is important to appreciate that as the organisms can only develop to infectious forms in ticks, the life cycle can be interrupted by taking steps to prevent ticks

Table 3.1 Some veterinary micro-organisms

Disease	Causative organism	Approximate size (nanometres)
Foot-and-mouth disease	virus	30
Contagious bovine pleuropneumonia	mycoplasma	150
Enzootic abortion of ewes	chlamydia	275
Anaplasmosis	rickettsia	300
Staphylococcal mastitis	bacterial coccus	1000
Anthrax	bacterial bacillus	3000–10,000 (length)
Babesiosis	apicomplexan protozoa	3000
Trypanosomosis	flagellate protozoa	25,000 (length)

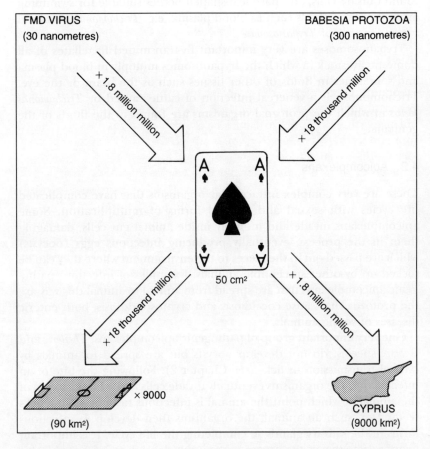

Fig 3.5 Relationships in size between *Babesia*, foot-and-mouth disease virus and a playing card.

attaching to and feeding off animals. Hence tick control is very important in the control of theileriosis and babesiosis.

5   How small are micro-organisms?

As mentioned above, micro-organisms are so small that they can only be seen with the aid of a microscope. By contrast, the arthropods and helminths outlined in Chapter 2 can be seen with the naked eye, although mites that parasitise the skin are just at the limit of visibility and usually require a microscope for detection.

But how small are micro-organisms? We can all imagine how large a mountain is in comparison to a house; the same sort of scale has to be used in the opposite direction for micro-organisms. It is conventional to measure micro-organisms in either micrometres, usually called microns for short, or nanometres. There are a million microns in a metre, and a thousand nanometres in a micron. Some of the micro-organisms are listed in Table 3.1.

Particles as small as 200 nanometres can be seen in conventional laboratory microscopes, so it can be seen from Table 3.1 that this is powerful enough for the detection of protozoa, bacteria, rickettsia, chlamydia and the larger mycoplasma. To detect viruses, however, much more powerful electron-microscopes are needed which can detect particles as small as just under half a nanometre.

In order to give some perspective to micro-organisms, they can be compared with everyday objects. Playing cards have an approximate area of about 50 cm^2, or nearly 1.8 million million times the size of a small virus like the foot-and-mouth disease virus. Multiplying a playing card by a similar factor gives you an area of about 9000 square kilometres, which is the approximate area of Cyprus. Applying a similar calculation to *Babesia*, one of the larger micro-organisms, gives you an approximate area of 9000 soccer pitches. These calculations are illustrated in Fig 3.5.

4 Recognition of diseases – signs of health

In the first three chapters we saw that there is a range of different types of diseases. In a disease outbreak, by observing the patterns by which the disease presents itself, we may be able to make an intelligent guess as to which type of disease it is. Further clues are the geographical location and the species of animal, and by referring to the disease distribution maps and Table 6.1 in Chapter 6, it may be possible to narrow down the possible diagnoses. How can we go beyond this stage and decide whether a sick animal is infected with a virus or harbouring a heavy worm burden? This is where skill, education and experience comes in. A qualified veterinary surgeon spends four or five years at a university acquiring the necessary skills and training to diagnose diseases, and even veterinary surgeons have to seek the help of specialists from time to time. Therefore it is unrealistic to expect others to be able to do the same but nevertheless, within certain limits, it is possible for lay people to diagnose diseases. Indeed, common animal diseases are diagnosed regularly by farmers the world over.

In any disease, the sick animal shows certain abnormalities. These abnormalities vary enormously and provide important clues for the veterinarian to make a diagnosis. Thus, an ox with rinderpest will have a range of fairly obvious abnormalities including diarrhoea, fever, depression and discharges from the eyes, nose and mouth, and farmers and veterinarians alike usually have little difficulty in suspecting this serious disease. These abnormalities are conventionally referred to as the clinical signs of disease and Chapter 5 outlines some of those which lay people could reasonably recognise. Before anyone attempts to identify clinical signs of disease in animals, however, they must know what healthy normal animals look like. Signs of health are just as important to appreciate as signs of disease, and these are considered in this chapter.

So what does a healthy animal look like? The diagnostic tables in Chapter 5 are designed round some of the more obvious clinical signs, e.g. death, diarrhoea and nervous signs. For the purpose of this book the corresponding 'normal signs' in a healthy animal are now briefly described.

1 Deaths

This may be regarded as a strange normal sign, but death is inevitable and it is important to appreciate that the occasional fatality in a flock or herd need not necessarily mean there is a serious disease outbreak. There is an old adage that in a herd or flock, one death is normal, two deaths are a coincidence and three deaths mean trouble and although mathematically minded epidemiologists may throw their hands up in horror at such a simplistic generalisation, it is surprising how often deaths stop after one or two animals.

So how do you recognise 'normal' deaths? Normal deaths can be attributed to old age, and the occasional accident or incident which may kill one or two animals at a time. Anything other than that can be considered as abnormal.

2 General condition

Some of the most important diseases do not cause dramatic clinical signs or kill significant numbers of animals, but affect animals' general condition and reduce their productivity. Diseases that cause dairy cattle to reduce their milk production, or interfere with liveweight gain of trade sheep, can seriously affect farmers' livelihoods just as much as a killing disease. In fact, such diseases can often be more economically damaging because their non-dramatic nature may result in failure by farmers to take any steps to control them. Thus it is important to appreciate whether animals are in optimal condition, because anything less than this may indicate an underlying disease problem that can be remedied.

Healthy animals in good condition have a sleek appearance to their coat. They should be well muscled so that the rib and pelvic bones are not prominent, but nicely rounded. The sides should continue in a fairly smooth convex line from behind the elbow to the front of the hind legs, and not be tucked up behind the ribs. These signs are readily visible in thin-skinned or light-coated animals, but less so in animals with heavy coats such as certain breeds of sheep. Experienced veterinarians and farmers automatically run their hands over animals when examining them to feel whether they are in good condition or not, and anyone involved with livestock should develop this expertise by constant practice (Fig 4.1).

Fig 4.1 Veterinarians examining animals automatically run their hands over them to feel the skin and assess the body condition.

2.1 Condition scoring

This can be taken a step further by condition scoring. This is a technique in which animals are given a score based on their body condition. To a certain extent the scoring is subjective and was originally developed for temperate breeds of cattle and sheep. The technique has now been adapted for use with tropical breeds of cattle, sheep and goats. The scoring scales vary depending on the technique used, e.g. 1 to 4 for goats, 0 to 5 for sheep and 1 to 9 for Boran cattle. At one end of the scale emaciated animals score low while those that are fat are given a high score. The author has found the technique very useful on numerous occasions in the tropics. By habitually condition scoring livestock, it is possible to assess very effectively whether they are losing or gaining condition in different circumstances, for example, after being treated for worms, grazing in different seasons, etc. (Fig 4.2). For an example of condition scoring, readers are referred to the book on sheep in this series.

3 *Skin*

We have already seen the importance of appreciating the sleek, smooth coat of a healthy animal in good condition. This can be taken a step further by a closer examination of the skin. By parting the wool or hair, the skin can be examined closely. It should be smooth and free of any lumps, loose scabs, flakes or debris.

The skin is a tough vital organ providing protection for underlying tissues and helping to maintain normal body temperature (see 'Heat stress' in Volume 2). It has to allow movement of underlying tissues (bones, muscles, tendons, etc.) and therefore healthy skin must be supple and flexible. If a fold of loose skin is picked up and then released, in a healthy animal it will immediately spring back to its original position. Failure to do so indicates some underlying problem with the circulation to the skin which can result from a number of causes, such as dehydration caused by diarrhoea.

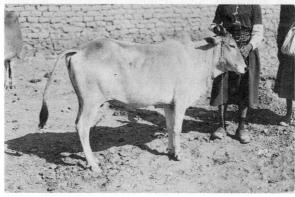

Fig 4.2   (above) Cow before treatment classified as in fair body condition (condition score 3).
(below) Same cow after 6 months phosphorus supplementation classified as in good body condition (condition score 7).

4 Head

Veterinarians examining animals can tell a great deal from the head. The eyes should be clear, bright and moist. Lachrymal glands secrete a permanent flow of tears which flow over the surface of the eyes, helping to protect them by washing away any particles or organisms that land on the eye surface. Special ducts in the corner of the eyes drain off this steady flow of tears into the back of the throat, so there should be no spillage onto the skin under the eyes.

The muzzle should be slightly moist and cool, and there should be no discharges from the nostrils. The inside of the mouth should always be wet from the continual secretion of saliva to aid digestion; the rate of secretion of saliva is such that it can be swallowed comfortably and not drool from the mouth.

5 Visible mucous membranes

Within an animal's body are a number of hollow or tubular organs which are lined by mucous membranes. These include the entire digestive tract from the mouth to the rectum, the respiratory tract from the nostrils to the lungs, the urinary and genital passages, and the conjunctiva which lines the inside of the eyelid and is reflected onto the front surface of the eyeball which it covers. The mucous membranes secrete a slimy substance called mucus which helps to lubricate the lining of these organs and provide a protective barrier by trapping foreign micro-organisms and particles which are then removed by various mechanisms. The mucous membranes are visible at certain points, namely the linings of the mouth (gums), nostrils, eyelids (conjunctiva), vagina, prepuce and rectum. In a healthy animal, they are smooth and glisten from the thin film of mucus. Where there is no pigment, they have a 'salmon pink' colour. In practice, the inside of the mouth and the conjunctiva are the most convenient visible mucous membranes to examine for signs of health or sickness.

NB The conjunctiva is transparent where it covers the front surface of the eyeball.

6 Nervous demeanour

Some important diseases affect the nervous system, and so it is important to appreciate normal animal behaviour. Domestic animals are usually accustomed to people and should have no fear of them. A stranger in their midst, however, will arouse their curiosity and attract their attention, even to the point of them approaching the stranger for a closer look. Any lack of such healthy curiosity may indicate that something is wrong.

Like people, some animals are 'bad tempered' and if confronted with an aggressive animal, it is important to ascertain whether this is normal or abnormal, as such behaviour can be an early sign of disease.

7 Gait

Lameness, whatever the cause, can be very distressing for an animal, and it is important to appreciate whether an animal's movement when walking (gait) is normal or not. Walking should be a balanced rhythmical motion, with the head swaying or nodding slightly in time with the animal's movement. When standing, the animal should be comfortable on all four feet and not show any excessive inclination to lie down or to lift any particular foot.

8 Respiration

In healthy animals, the movement of breathing in and out is almost silent and barely noticeable except after physical exertion. The number of times an animal breathes in and out in a minute when resting is called the respiration rate, and this varies from breed to breed as well as species to species. As a general rule, the larger the animal, the slower the respiration rate. The following can be used as a rough guide of rates for healthy animals at rest:

Horses	8–12	Cattle	12–16
Camels	6–12	Sheep/goats	12–20
Buffaloes	21–24	Pigs	10–16

Respiration has three equal phases, breathing in, breathing out and a pause. The respiration rate can increase in a healthy animal following exercise or an increase in environmental temperature or humidity.

9 Body temperature

The health of all animals depends on their ability to maintain almost constant body temperatures even if subjected to extremes of environmental temperatures. They can do this by a variety of physiological and behavioural mechanisms (see 'Heat stress' in Volume 2), and significant deviation from the normal range of body temperature is a strong indication of ill health.

Hence it is not surprising that veterinarians attach considerable importance to animals' body temperature when examining them. Body temperature is measured by placing the bulb of a clinical thermometer inside the rectum and holding it in position for about a minute (digital electronic thermometers are now available which give a faster reading). Needless to say the thermometer should be cleaned with cold disinfectant after every use to avoid the risk of transferring pathogenic organisms from one animal to another.

NB Do not use hot water to clean the thermometer as this will damage it.

As with respiration rates, the normal range of body temperature varies from species to species and, to a certain degree, breed to breed. Temperatures also show diurnal variation, being lowest early in the morning and highest in late afternoon. Camels in particular may show fluctuations of 6°C or more in their body temperature, a characteristic which allows them to cope with extreme environmental temperatures and conserve water which would otherwise be lost by evaporation as a physiological cooling mechanism. As a general rule the larger the animal, the lower its body temperature, and for most practical purposes, the following can be used as a rough guide to average body temperatures of healthy animals:

	°F	°C		°F	°C
Camels	99.5	37.5	Pigs	102.0	39.0
Buffaloes	101.0	38.3	Sheep	102.0	39.0
Cattle	101.5	38.5	Goats	103.0	39.5
Horses	100.5	38.0			

10 Faeces

The consistency and colour of faeces of healthy animals obviously depends on their diet. Consequently when working with domestic animals it is important to appreciate this and recognise 'normal' faeces of healthy animals on different diets, grazing, types of management, etc. Young sucking animals produce yellowish pasty faeces which change in colour and consistency as they change from milk to solid food.

Adult cattle on a diet of mainly green grass pass fairly wet faeces as often as 18 to 20 times a day, whereas cattle extensively reared on semi-arid rangeland pass drier faeces and defecate less frequently. Buffaloes also pass large quantities of faeces, most of which is moisture. Camels, horses, sheep and goats pass pelleted faeces. Those of camels are oblong and about 4 cm in length whereas those of horses are more irregular in shape and about 4 – 8 cm across. The pelleted faeces of sheep and goats are much smaller and rounder. The consistency of faeces of all healthy animals can change very quickly with a change of diet. Diarrhoea is an important clinical sign of many diseases but its presence may indicate no more than a flush of green grazing after rains or a change of pasture.

11 Reproduction

Some major diseases disrupt reproduction, for example, by causing abortions, infertility, etc. A detailed description of the reproduction cycles of domestic animals is beyond the scope of this book but the important features are summarised in Table 4.1.

11.1 Oestrus

Oestrus is the period when the sexually mature female accepts the male. This is mainly controlled by complex internal hormonal changes, but there are several extraneous factors that also influence oestrus. In temperate climates with considerable variations in day length between summer and winter, changes in the length of daylight affect sexual cycles of domestic livestock, particularly sheep and goats. Thus sheep come into oestrus and breed when there is a shortening day-length (winter) and lamb in spring. With relatively little variation in day-length in tropical and subtropical regions, other factors are more significant, e.g. high ambient temperatures or low nutritional status are liable to suppress sexual activity. Breeding seasons thus vary considerably from region to region, and can only be appreciated by acquiring local knowledge. During oestrus the adult female ovulates and if she is not successfully mated by a male, she will return to oestrus again. This period is referred to as the oestrus cycle and although it varies considerably from animal to animal and from breed to breed, the averages and ranges shown in Table 4.1 can be used as a guide for most practical purposes.

Camel cows do not come into oestrus and ovulate spontaneously like other domestic animals, but are stimulated to ovulate by the act of copulation. The term 'oestrus cycle' thus strictly speaking does not apply to camels, and the figure in Table 4.1 refers to the time that passes between one period of acceptance of the male and the next.

Table 4.1 Reproduction data of domestic animals

	Gestation period (days)	Oestrus cycle (days) Average	Range
Cattle	279–292	21	18–24
Buffaloes (River)	302–319	21	11–30
Buffaloes (Swamp)	301–343	21	11–30
Camels (Dromedary)	336–405	24*	15–28
Sheep	140–160	17	14–21
Goats	145–155	19	18–21
Pigs	110–117	21	16–30
Horses	330–342	21	19–26

*Camels stimulated to ovulate by the bulls; see text.

11.2 Gestation period

The gestation period is the duration of pregnancy and as can be seen from Table 4.1, each species has a considerable range due mainly to breed variations.

The signs of health outlined above are summarised in Fig 4.3. These should only be regarded as an approximate indication of what to look for in a healthy animal as there is no substitute for practical experience in acquiring the skill and knowledge to do the job properly.

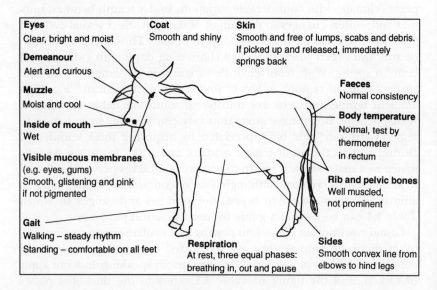

Eyes
Clear, bright and moist

Demeanour
Alert and curious

Muzzle
Moist and cool

Inside of mouth
Wet

Visible mucous membranes
(e.g. eyes, gums)
Smooth, glistening and pink
if not pigmented

Gait
Walking – steady rhythm
Standing – comfortable on all feet

Coat
Smooth and shiny

Skin
Smooth and free of lumps, scabs and debris.
If picked up and released, immediately
springs back

Faeces
Normal consistency

Body temperature
Normal, test by
thermometer
in rectum

Rib and pelvic bones
Well muscled,
not prominent

Respiration
At rest, three equal phases:
breathing in, out and pause

Sides
Smooth convex line from
elbows to hind legs

Fig 4.3 Signs of health in animals.

5 Recognition of diseases – clinical signs

With an appreciation of the signs of good health in animals outlined in Chapter 4, one is then in a position to recognise the signs of diseases or clinical signs. Veterinarians are professionally trained to do this, but farmers from experience are also experts at recognising clinical signs of disease. They may not know what the problem is but they are quick to recognise that something is wrong.

We saw in the first chapter that by observing how a disease presents itself, it may be possible to assess which kind of disease it is. For most farmers, however, the clinical signs of disease are the most obvious feature. It will be more apparent to him if the disease affecting his animals causes diarrhoea or coughing or deaths, etc. rather than whether it is infectious or arthropod-borne.

The following brief descriptions outline the signs of sickness that are seen in the main diseases of animals in tropical countries. They are then followed by a series of diagnostic tables based on the main presenting clinical signs of the diseases described in Volume 2. Some diseases have more than one prominent clinical sign and thus appear in more than one diagnostic table. For example, rinderpest is a frequently fatal disease in which affected animals usually have severe diarrhoea before dying and so rinderpest appears in tables of killing diseases as well as in tables for diarrhoeic diseases. As for the signs of health, they can be used as a guide and may be helpful when professional veterinary help is not available, but it must be stressed that this chapter should not be regarded as some sort of DIY veterinary manual.

1 Deaths

As mentioned above, death is inevitable and only when abnormal levels of deaths occur in a flock or herd can this be regarded as a sign that a disease is present. It could be argued that any disease has the potential to be fatal. Old people in failing health are liable to die from a minor ailment such as the common cold, but it would be foolish to regard the cold as a killing disease. The same applies to animals and the killing diseases referred to in the tables below are diseases in which there is a high probability of death, particularly if animals are not treated.

An important characteristic of killing diseases is whether animals die after a short illness or not because this can be an important diagnostic pointer. The tables thus differentiate diseases that result in rapid death. This feature is often referred to as sudden death but in reality there are very few diseases that cause animals to drop dead without first being ill. In reality sudden death usually means that an animal has died after a very short illness which went undetected and has just been found dead.

NB Books on diseases often quote the morbidity and mortality rates of different diseases. The morbidity rate of a disease is the percentage of a population at risk that become sick with the disease in an outbreak. The case mortality rate is the percentage of sick animals that die from the disease whereas the population mortality rate is the percentage of animals at risk that die from the disease. I have deliberately avoided these definitions because they can be confusing and, because these rates can vary a great deal, any figures quoted are liable to be inaccurate in many situations. Important killing diseases are presented in Tables 5.1 to 5.4.

2 *Loss of condition*

Some of the most important diseases of livestock in the tropics do not cause dramatic clinical signs but cause progressive loss of condition which, if not reversed, can ultimately be fatal. Animals in poor or failing condition lack the alertness and energy of healthy ones. Their coats are liable to be dull and coarse, rather than smooth and sleek. The most obvious sign is loss of body condition; the smooth rounded curves gradually disappear and are replaced by obvious protrusions of poorly fleshed bones showing through the skin; rib and pelvic bones are particularly prominent. The abdomen, instead of running in a smooth curve from behind the rib cage, becomes tucked up. These signs are fairly obvious, even to the untrained eye, but can be measured by condition scoring as outlined in Chapter 4. Diseases which cause loss of condition in ruminants are presented in Tables 5.5 and 5.6. As many of these diseases also cause diarrhoea, the tables indicate those in which diarrhoea is likely to be a clinical sign.

Table 5.1 Killing diseases of cattle and buffaloes

	Rapid death	Epidemiological picture	Vol. 2 page
anthrax	★	Disease of grazing ruminants; high fever and death, often sudden; blood discharges from external orifices after death	35
babesiosis (*B. bovis*)		Tick-borne; cattle from non-endemic areas very susceptible; fever, anaemia, red-water and nervous signs; sometimes high death rates	135
bacillary haemo-globinuria	★	Enterotoxaemia of cattle with liver fluke (fasciolosis); fever, abdominal pain, red coloured urine and death	167
black disease		Also enterotoxaemia of cattle with liver fluke; depression and death after 1 or 2 days	167
black quarter	★	Disease of young grazing cattle; hot painful swellings of hind- and fore-quarters, lameness, fever and death in 1–2 days	167
botulism		Poisoning from organic material; paralysis, tongue protrusion, recumbency and death after 1–2 weeks; occasional rapid collapse and death	205
cyanide poisoning	★	Poisoning from young/rapid growing plants, e.g. sorghum; red mucous membranes, difficult breathing, collapse and death	198
D. cymosum poisoning	★	Leaves contain heart toxin which causes acute circulatory failure, collapse and death, often after drinking water	200
fat cow syndrome		Uncommon; dairy cows overfed when dry; after calving, loss of appetite, coma and death	180
foot-and-mouth disease	★	Very contagious; causes sudden deaths in calves	47
gousiekte	★	Accumulative poisoning by plants which cause fibrosis of heart; sudden death from heart failure 1–2 months later, often after exercise	201
heartwater		Tick-borne; cattle from non-endemic areas very susceptible; fever, nervous signs and often death	129
heat stress		Overexposure to high ambient temperatures or overcrowding with poor ventilation; lethargy, hyperthermia, collapse and death	193
haemorrhagic septicaemia		Major disease especially of buffaloes; fever, salivation, nasal discharges and death in 1–2 days; usually after stress e.g. start of ploughing	4
malignant catarrhal fever		Sporadic; cattle in contact with calving wildebeest or lambing sheep; eye/nose discharges, enlarged head/neck lymph nodes and fever	6

Table 5.1 – *cont'd*

	Rapid death	Epidemiological picture	Vol. 2 page
mucosal disease		Sporadic disease of cattle; clinical signs resemble rinderpest; usually fatal	76
rabies		Usually infected by bite of another rabid animal; e.g. dog, jackal, etc; nervous signs for about a week before death	58
Rift Valley fever		Mosquito-borne; acute disease in calves; fever, vomiting, nasal discharges, collapse and death; peracute disease in newborn	108
rinderpest		Contagious major killing disease; severe diarrhoea, eye/nose discharges, salivation, fetid odour, death about 10 days (usually) later	60
seneciosis (acute)	★	Rapid ingestion of large quantities of *Senecio* weeds, e.g. as young shoots after range fire; internal haemorrhages and sudden death	199
theileriosis		Tick-borne disease; exotic animals very susceptible and may suffer high death rates; fever, anaemia, jaundice and enlarged lymph nodes	131

Table 5.2 Killing diseases of sheep and goats

	Rapid death	Epidemiological picture	Vol. 2 page
acute fasciolosis		Heavy infection over short period from grazing aquatic snail habitat; weak, anaemia and death after 1–2 days; usually young sheep	151
anthrax	★	Disease of grazing ruminants; high fever and death, often sudden; blood discharges from external orifices after death	35
bacillary haemo-globinuria	★	Enterotoxaemia of sheep with liver fluke (fasciolosis); fever, abdominal pain, red coloured urine and death	167
black disease		Another enterotoxaemia; affects sheep with liver fluke; depression and death after 1 or 2 days	167
black quarter	★	Occasional disease of young grazing sheep; hot painful swellings of hind- and fore-quarters, lameness, fever and death in 1–2 days	167
botulism		Poisoning from organic material; paralysis, tongue protrusion, recumbency and death after 1–2 weeks; occasional rapid collapse and death	205

Table 5.2 – *cont'd*

	Rapid death	Epidemiological picture	Vol. 2 page
CCPP*		Major contagious lung disease of goats; fever, laboured breathing, cough and nasal discharges; 60–90% can die in an outbreak	12
cyanide poisoning	★	Poisoning from young/rapid growing plants, e.g. sorghum; red mucous membranes, difficult breathing, collapse and death	198
D. cymosum poisoning	★	Leaves contain heart toxin which causes acute circulatory failure, collapse and death, often after drinking water	200
entero-toxaemia	★	Usually predisposed by change in diet; fever, depression, collapse and death	167
gousiekte	★	Accumulative poisoning by plants which cause fibrosis of heart; sudden death from heart failure 1–2 months later, often after exercise	201
haemonchosis		Very important in grazing sheep; lambs especially susceptible; anaemia, loss of condition and death in severe cases	143
heartwater		Tick-borne; sheep/goats from non-endemic areas very susceptible; fever, nervous signs and often death	129
Nairobi sheep disease		Tick-borne; non-indigenous animals very susceptible; fever, eye and nose discharges, abortions, dysentery, collapse and death	130
PPR**		Very contagious disease of goats and sometimes sheep; severe diarrhoea, eye/nose discharges, salivation and death	20
pregnancy toxaemia		Ewes in late pregnancy; early signs vague; listlessness, change in behaviour, nervous signs, sickly sweet smell and death in a few days	181
rabies		Rare in small ruminants; nervous signs for about a week before death	58
Rift Valley fever		Mosquito-borne; acute disease in lamb/kids; fever, vomiting, nasal discharges, collapse and death; peracute disease in newborn	108
rinderpest		Contagious major killing disease of cattle; virus strains in India pathogenic to sheep and goats; disease very similar to PPR	60
sheep and goat pox		Widespread; all ages susceptible; fever and pox lesions in thin skinned areas; occasional deaths	25

Table 5.2 – *cont'd*

	Rapid death	Epidemiological picture	Vol. 2 page
struck	★	Mainly in sheep up to 1 year old; often follows improvement in nutrition; sudden onset, unsteady gait, convulsions and death in a few hours	167
theileriosis		Tick-borne disease; exotic breeds very susceptible; fever, enlarged superficial lymph nodes and death	131

*contagious caprine pleuropneumonia   **peste des petits ruminants

Table 5.3   Killing diseases of pigs

	Rapid death	Epidemiological picture	Vol. 2 page
aflatoxicosis		Poisoning from fungal contamination of stored feed; young pigs particularly susceptible; unthriftiness and death after short illness	202
African swine fever	★	Spread by soft ticks and direct contact; main signs are fever, incoordination and death after a week; can be peracute or, rarely, chronic	27
anthrax		High fever, oedematous swellings of throat and death after 2–3 days; blood discharges from external orifices after death	35
cotton seed cake poisoning		Gossypol poisoning from diets with >10% cotton seed cake; prolonged feeding causes heart damage and eventual heart failure	202
entero-toxaemia	★	Uncommon in tropics; affects sucking piglets up to 1 week old; depression, diarrhoea with blood and death within 24 hours	53
erysipelas (acute)		Common infection; all ages susceptible; fever, eye discharges, red diamond shaped skin swellings and death, in many, 3 days later	46
erysipelas (chronic)	★	Sudden death from heart lesions; may be normal or show signs of circulatory failure, e.g. laboured breathing, blue skin colouration	46
foot-and-mouth disease	★	Very contagious; causes sudden death in young piglets	47
heat stress		Can occur if pigs denied access to shade and water; lethargy, collapse and death	193

Table 5.3 – *cont'd*

	Rapid death	Epidemiological picture	Vol. 2 page
hog cholera		Common and very infectious; resembles acute (commonest), sub-acute and chronic forms of African swine fever	28
mulberry heart disease	★	Rapidly growing young pigs on high energy concentrates deficient in selenium or Vitamin E; heart lesions cause heart failure and sudden death	189
rinderpest		Contagious; salivation, eye/nose discharges, diarrhoea and death about 10 days (usually) later	60
Trypanosoma simiae	★	Transmitted by tsetse flies; high fever and death within 24 hours of clinical signs appearing	121

Table 5.4   Killing diseases of horses and donkeys

	Rapid death	Epidemiological picture	Vol. 2 page
African horse sickness		Major killing disease; transmitted by midges; 2 main forms, acute pulmonary and subacute circulatory; mixed form and milder form also	111
anthrax		Major killing disease; high fever, oedematous swellings of throat and death after 2–3 days; blood discharges from external orifices after death	35
babesiosis (redwater)		Tick-borne disease; fever, anaemia, jaundice and haemoglobinuria; ranges from subclinical, to severe with death in 1–2 days of first signs	135
botulism	★	Bacterial poison from organic matter; progressive paralysis, tongue protrusion and death in 1–2 weeks; sometimes peracute and rapid death	205
dourine		Sporadic; chronic venereal disease with lesions of genitalia, skin oedema and loss of condition; high death rate in clinical cases	70
EE*		Mosquito-borne infections; subclinical to severe (fever, nervous signs, collapse and death); up to 80% of American infections may die	114
equine inf. anaemia		Spread by biting flies; fever, eye/nose discharges and oedema; some die in a few weeks but many recover only to relapse and die later	113
heat stress		Overexposure to high ambient temperatures or overcrowding with poor ventilation; lethargy, hyperthermia, collapse and death	193

Table 5.4 – *cont'd*

	Rapid death	Epidemiological picture	Vol. 2 page
rabies		Uncommon; usually infected by bite of another infected (rabid) animal, e.g. dog/jackal; nervous signs and death in about a week	58
seneciosis (acute)	★	Rare; rapid ingestion of large amount of *Senecio* weeds e.g. young shoots after range fire; internal haemorrhages and sudden death	199
strongylosis *(S. vulgaris)*		Rare; larvae block internal arteries and resultant tissue damage causes fever, abdominal pain and sometimes death	161
tetanus		Horses very susceptible; contamination of cuts/ wounds etc.; stiffness, whole-body spasms, convulsions and death from respiratory failure	167
trypanosomosis		Fly-borne infections; cause chronic anaemia, loss of condition and often eventual death, sometimes many months after first clinical signs	121

*Equine encephalomyelitides; (Eastern EE; Western EE; Venezuelan EE; Japanese EE; West Nile EE)

Table 5.5   **Diseases of cattle and buffaloes characterised by loss of condition**

	Diarr- hoea	Epidemiological picture	Vol. 2 page
acetonaemia		Affects high yielding dairy cows a few weeks after calving; smell of acetone in breath and milk; milk yield drops; most recover	180
aflatoxicosis		Affects calves; ingestion of fungus contaminated grain or other fodder causes liver damage; may die after a few weeks	202
ascariosis	★	Affects calves, especially buffaloes, in first 6 months; usually infected from mothers via the milk when sucking; sometimes fatal	150
Ca/P* deficiency		Ca deficiency rare but P deficiency common in grazing cattle; limb deformities in young; stiff- ness, bone fragility, loss of appetite in adults	186
fasciolosis	★	Affects cattle of all ages; infection from grazing near aquatic snail habitat; signs may include anaemia, and 'bottlejaw'; can be fatal	151
cobalt deficiency		Affects grazing ruminants in many parts of the world; can vary from mild to severe emaciation and death	187

Table 5.5 – *cont'd*

	Diarr-hoea	Epidemiological picture	Vol. 2 page
copper deficiency	★	Affects grazing ruminants in many areas; clinical signs include anaemia, poor growth, loss of hair pigment and diarrhoea	187
Dicro-coelium inf.	★	Uncommon disease similar to fasciolosis	156
Johne's disease	★	Sporadic disease; uncommon in tropics; affects adults; chronic wasting disease with foul smelling diarrhoea; usually fatal	50
malnutrition		Usually due to overgrazing following drought, etc.; may be complicated by other conditions, e.g. internal parasites	184
PGE** (worms)	★	Major problem of ruminants, particularly on heavily grazed pastures; all ages susceptible but young usually worst affected	143
rinderpest	★	Contagious and highly fatal disease; fever, diarrhoea, eye/nose discharges and salivation; significant loss of condition before death	60
schistosomosis	★	Occurs in cattle grazing near aquatic snail habitat	154
seneciosis	★	Grazing poisonous *Senecio* weeds; usually chronic disease due to liver damage; acute poisoning and death sometimes	199
sodium deficiency		Sodium deficient pastures widespread, especially common in Africa; craving for salt causes urine drinking, licking of soil, coats, etc.	187
trypanosomosis		Major chronic wasting condition; transmitted to cattle by tsetse flies in sub-Saharan Africa and by biting flies to buffaloes in Asia	121
tick infestation		Excessive tick burdens a common cause of loss of condition	93
tuberculosis		Major debilitating infection of cattle; lung involvement common resulting in breathing difficulties and chronic cough	66

*calcium/phosphorous deficiency   **parasitic gastro-enteritis

Table 5.6 Diseases of sheep and goats characterised by loss of condition

	Diarr-hoea	Epidemiological picture	Vol. 2 page
blue-tongue	★	Midge-borne; widespread; indigenous ruminants rarely affected; exotics may suffer severe disease; fever, eye/nose/mouth discharges and lameness	104
cobalt deficiency		Affects grazing ruminants in many parts of the world; can vary	187
coccidiosis	★	Affects young animals in crowded conditions; blood/mucus in diarrhoea; rarely fatal; prolonged convalescence in severe cases can impair growth	41
copper deficiency	★	Affects grazing ruminants in many parts of the world; clinical signs include anaemia, poor growth, and chronic diarrhoea	187
Dicrocoelium infection	★	Uncommon; occurs in animals grazing near aquatic snail habitat	156
fasciolosis	★	All ages susceptible; infection from grazing aquatic snail habitat; anaemia and 'bottlejaw', can be fatal; sheep, being grazers, at greater risk	151
jaagsiekte		Slow progressive infection of sheep causing lung tumours; increasingly laboured breathing and death after several months; +ve 'wheelbarrow' test	15
Johne's disease	★	Sporadic disease; uncommon in tropics; affects adults; chronic wasting disease with foul smelling diarrhoea; usually fatal	50
maedi		Slow progressive fatal infection of adult sheep and sometimes goats; clinical signs similar to jaagsiekte but 'wheelbarrow' test is negative	16
malnutrition		Usually due to overgrazing following drought, etc.; often complicated by other conditions, e.g. internal parasites; sheep more at risk than goats	184
mange		Mange mite infestation of skin can cause intense irritation and debilitation; sheep particularly susceptible to *Psoroptes*, cause of sheep scab	101
pediculosis (lice)		Infestation of skin with lice; heavy burdens cause irritation and stress; associated with unhealthy animals in overcrowded conditions	91
PGE* (worms)	★	Major problem of grazing ruminants, particularly on heavily grazed pastures; all ages susceptible especially young; sheep at greater risk	143

Table 5.6 – *cont'd*

	Diarr-hoea	Epidemiological picture	Vol. 2 page
schistosomosis	★	Uncommon; occurs in ruminants grazing near aquatic snail habitat; diarrhoea may contain blood and mucus	154
scrapie		Progressive fatal infection of adult sheep and occasionally goats; slight nervous signs increasing over several months plus intense skin irritation	23
sheep and goat pox		Widespread; all ages susceptible but young worst affected; fever, eye/nose/mouth discharges and pox lesions in thin haired areas; can be fatal	25
tuberculosis		Debilitating infection; sheep fairly resistant but goats sometimes affected; lungs commonly affected causing breathing difficulties and chronic cough	66
trypanosomosis		Transmitted by tsetse flies; fairly uncommon in small ruminants; chronic anaemia, loss of condition and sometimes death	121
visna		Same infection as maedi; virus infects brain causing nervous signs, incoordination, paralysis and death; can affect sheep from 2 years of age	16

*parasitic gastro-enteritis

3 Skin lesions

As outlined in the previous chapter, the skin of healthy animals is smooth and should contain no lumps in or under the skin. Livestock in the tropics are commonly at risk from skin arthropods (see Chapter 3), some of which cause irritation. Skin irritation is usually fairly obvious; animals attempt to gain relief by rubbing against objects or nibbling their skin (Fig 5.1). Areas of reddened skin with loss of hair/wool should be regarded as a suspicious sign of rubbing due to irritation.

A description of skin lesions of different diseases could fill a book itself and, in any case, requires a great deal of skill and experience to diagnose. For the purpose of this book, skin lesions can be regarded as lumps or nodules (either in or under the skin), scabs, oedema and thickening.

Extensive skin lesions over the body are usually obvious, but some diseases cause lesions in the skin of the extremities, i.e. tips of ears, nose, lips, tail, feet and lower limbs. These may be missed unless the animal is examined carefully. Diseases with characteristic skin lesions are presented in Tables 5.7 to 5.10 (pp. 56–61).

Fig 5.1   Sheep with scrapie; the sheep is attempting to gain relief from
intense skin irritation by rubbing itself against the pen (CTVM).

3.1   Lumps and nodules

Lumps and nodules (small lumps) may be visible depending on their
size and number, and on the thickness of the coat. A visual inspection
alone is not enough, however, and anyone working with livestock should
regularly run their hands over their animals as this practice will detect
any lumps hidden under the coat. Once lumps have been detected, the
next thing is to determine by palpation and lifting up the skin whether
they are part of the skin or underneath it.

Lumps or nodules under the skin may indicate reactions in the lymphatic system. At the heart of an animal's complex array of mechanisms
to defend itself against infectious micro-organisms is the lymphatic
system which comprises a network of lymph nodes and vessels circulating
lymphocytes, vital immune cells, through the blood stream and tissues.
This network is found throughout the body and those lymph nodes
that lie under the skin are sometimes called superficial lymph nodes
(Fig 5.2). Some infections may result in so much reaction that the
lymph nodes are visibly enlarged and can be palpated. Reactions in the
linking lymph vessels may cause inflammation and thickening

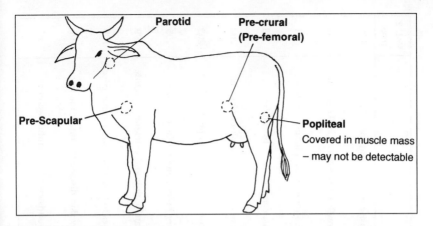

Fig 5.2   Superficial lymph nodes just under the skin which may become enlarged in certain diseases.

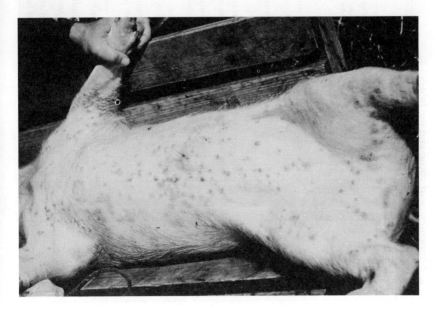

Fig 5.3   Prominent skin nodules of pig pox (CTVM).

(lymphangitis, see Figs 1.3 and 1.9, Volume 2) which results in 'cording' under the skin. Of course there is a range of other lesions that can cause lumps under the skin, such as abscesses, tumours, etc. Other diseases cause nodules in the skin itself, such as lumpy skin disease (Fig 4.1, Volume 2) and pig pox (Fig 5.3).

Table 5.7 Diseases of skin and appendages of cattle and buffaloes

	Skin lesions				Epidemiological picture	Vol. 2 page
	*Irrit.	Nod.	Sca.	Sep.		
besnoitiosis		★		★	Sporadic; mild (a few nodules) to severe (extensive nodules and sometimes death); lesions on scrotum and eyes common	119
blowfly myiasis		★		★	Uncommon; infestation of wounds, etc. with blowfly larvae; causes strike which can be extremely distressing	81
bovine farcy		★			Common; small sub-cutaneous nodules in neck, shoulders and limbs; painless and rarely causes ill-health; sometimes rupture	1
dermatophilosis			★		Problem in susceptible cattle in presence of *Amblyomma* ticks; scabs can be extensive with loss of condition	43
ergotism			★		Mycotoxicosis; usually from rye or other cereal; necrosis of extremities (feet, lips, tips of ears and tail); causes lameness	202
facial eczema			★		Occasional mycotoxicosis from contaminated pasture litter; causes photo-sensitisation similar to *Lantana* poisoning	203
horn cancer				★	Usually in Hariana castrates; occasionally in other breeds and buffaloes; horn drops off due to cancer of horn core	195
Lantana poisoning			★		Garden shrub now wild; photosensitisation, inflammation and cracking of lightly pigmented skin; can be severe with jaundice	199
lumpy skin disease (LSD)		★			Fly-born infection; fever and eye/nasal discharges followed by nodules; may cover whole body; debilitating but rarely fatal	106
LSD (pseudo)		★			Fly-born infection; clinical signs similar to lumpy skin disease but much milder; usually harmless and benign infection	106
malignant oedema				★	Uncommon; infection of contaminated cuts, wounds etc.; hot painful swellings, putrid smell (gas gangrene); often fatal	167

Disease	Signs and significance	Irritation	Nodules	Scab formation	Septic	Page
mange	Intense irritation from skin mites; usually in head, neck and legs, usually associated with crowding and poor husbandry	★				101
mange (demodectic)	Specific type of mange; causes nodular thickening in head, neck and forelimbs; unlike other forms, no irritation		★	★		101
pediculosis (lice)	Infestation with lice; heavy burdens cause irritation and stress; associated with unhealthy cattle in overcrowded conditions	★	★			92
ringworm	Fungal infection; raised greyish, circular scabs, usually in head and neck; usually associated with overcrowding			★	★	63
saddle sores	Lesions on body from badly fitting harnesses, saddles, etc.; raw areas of skin can become infected and painful				★	176
screw worm	Screw-shaped fly larvae burrow into skin wounds; heavy infestations cause large foul-smelling lesions			★	★	85
stephanofilarosis	Fly-born worm infestation; small lumps which discharge and haemorrhage over a few months causing thickening of skin		★		★	109
sweating sickness	Tick toxicosis of calves; fever, salivation, eye/nasal discharges and extensive moist skin eczema; can be severe and fatal				★	139
torsalo	Fly larvae burrow to sub-cutaneous tissues; painful swellings cause loss of condition and hide damage		★		★	85
urticaria	Acute allergic reaction; sudden plaque like swellings on any part of the body; usually disappear in a few hours		★			This vol.
warble fly	Fly larvae migrating through tissues eventually emerge through holes punctured in the skin; causes hide damage		★		★	85
worm nodule disease	Small worms spread by flies and midges; microfilarial larvae cause small benign subcutaneous nodules		★			126

*Irritation; nodules; scab formation; septic

Table 5.8 Diseases of skin and appendages of sheep and goats

	Epidemiological picture	Vol. 2 page
besnoitiosis	Uncommon; goats in Kenya; cysts on ears, eyes and genitalia	119
big head	Young rams; infection of damaged tissues in head from fighting; swelling of head, especially round eyes, ears, nose and lower jaw	167
blowfly strike	Infestation of wounds, faeces soiled fleece, etc. with blowfly larvae; can cause foul smelling lesions; mainly a problem of sheep	81
blue-tongue	Widespread; severe disease in exotic breeds of sheep; fever, discharges from eyes/nose/mouth, lameness and broken fleece	104
Border disease	Uncommon; congenitally infected lambs have hairy pigmented coats and uncontrollable jerky movements (hairy shakers)	75
caseous lymphadenitis	Common; chronic abscess infection of lymph nodes; superficial nodes under skin may rupture and discharge thick, green pus	7
dermatophilosis	Widespread; in sheep scabs under fleece hidden, but any on face and ears visible; in goats lesions on lips, muzzle, feet and scrotum	43
ergotism	Mycotoxicosis; usually from rye or other cereal; necrosis of extremities (feet, lips, tips of ears and tail); causes lameness	202
facial eczema	Photosensitisation from mycotoxicosis from pasture litter in intensive sheep; inflammation and cracking of skin, especially around head	203
malignant oedema	Uncommon; infection of contaminated cuts, wounds, etc.; hot painful swellings, putrid smell (gas gangrene); often fatal	167
mange	Mite infestation of skin can cause intense irritation and debilitation; sheep particularly susceptible to *Psoroptes*, cause of sheep scab	101
demodectic mange	Specific type of mange; nodular thickening of skin of goats on head, neck and forelegs; no irritation unlike other types of mange	101
orf	Common; raw, painful, bleeding scabs, especially on head and mouth (lambs and kids), feet (older animals), or teats of suckling adults	18
pediculosis (lice)	Infestation of skin with lice; heavy burdens cause irritation and stress; associated with unhealthy animals in overcrowded conditions	92

Table 5.8 – *cont'd*

	Epidemiological picture	Vol. 2 page
photosensitisation	Liver toxin from ingesting Devil's thorn; inflammation and cracking of lightly pigmented skin; can be severe with jaundice	200
ringworm	Uncommon; fungal infection; raised greyish, circular scabs, usually in head and neck; usually associated with overcrowding	63
sheep and goat pox	Widespread; all ages susceptible but young worst affected; fever, and pox lesions in thin haired areas; can be fatal	25

Table 5.9   Diseases of skin and appendages of pigs

	Epidemiological picture	Vol. 2 page
African swine fever	Major infectious disease; in acute form clinical signs include fever, incoordination and red skin blotches, particularly on the extremities	27
ergotism	Uncommon; poisoning from fungal contamination, usually of cereals or rye grasses; necrosis of extremities (feet, tips of ears and tail, etc.)	202
erysipelas (acute)	Common; characteristic red, often diamond shaped, skin swellings in ears, neck, lower abdomen and inner thigh	46
erysipelas (chronic)	Clinical signs more vague than acute form; may include loss of hair, skin thickening and sloughing of ear and tail tips	46
foot-and-mouth disease	Very infectious; clinical signs may include vesicle formation on the snout	47
hog cholera	Major infectious disease clinically similar to African swine fever	28
malignant oedema	Uncommon but often fatal; infection from contamination of skin wounds, syringe needles, etc.; hot painful swellings with putrid smell	167
pediculosis (lice)	Common in housed or closely confined pigs; often tolerated but heavy burdens can cause skin irritation and scratching	92
pig pox	Benign skin pox transmitted by lice and stable flies; usually in sucking and newly-weaned pigs; lesions on skin and in mouth; usually recover in 5–6 weeks	This vol.
ringworm	Uncommon; usually associated with prolonged over-crowding; ring shaped scabby lesions, usually on the back and sides	63
sarcoptic mange	Skin irritation and thickening; mainly ears but can spread to other parts of body; carrier sows common source of infection to sucking piglets	101

Table 5.10 Diseases of skin and appendages of horses and donkeys

	Skin lesions				Epidemiological picture	Vol. 2 page
	*Irrit.	Nod.	Sca.	Oed.		
besnoitiosis		★			Uncommon; life cycle unknown; parasitic cysts in skin, eyes and scrotum; usually mild disease	119
brucellosis					Rare; *Brucella* infection over point of shoulders (fistulous withers) or behind head; inflammation, rupture and discharges	38
dermatophilosis			★		Major cattle disease; affects horses also; scabs on head and lower limbs; extensively over body in severe cases	43
dourine				★	Venereal disease; oedema of genitalia can extend along abdomen; oedematous plaques may develop on body	70
equine infectious anaemia				★	Spread by biting flies; fever, eye/nose discharges and oedema on abdomen, legs and prepuce in stallions	113
epizootic lymphangitis		★			Infection tracking along lymph vessels in head, neck, shoulders and legs; nodules and discharging abscesses	31
glanders (skin form)		★			Uncommon; usually in chronic infection; painful nodules rupture leaving scars; any site but usually on limbs	32
mange	★				Fairly uncommon; mites infest skin in head, neck and legs; usually associated with crowding and poor husbandry	101
pediculosis (lice)	★				Heavy burdens cause irritation and stress; usually associated with overcrowding of unhealthy animals	92
ringworm			★		Fairly uncommon; raised greyish, circular scabs in head and neck usually; associated with overcrowding	63

60

	Irritation	nodules	scab formation	oedema under the skin*		
saddle sores		★			Lesions from badly fitting harnesses, saddles, etc.; raw areas of skin can become infected and painful	176
summer sores		★		★	Worm larvae deposited by flies around eyes, lips, nose and wounds causing lesions which scab over and ulcerate	117
sweet itch	★	★			Reaction to *Culicoides* midges; irritation with moist discharge and scabs of mane, tail and under abdomen	This vol.
trypanosomosis			★		Fly-borne infection; chronic anaemia and loss of condition; parasites may invade skin tissues causing oedema	121
ulcerative lymphangitis				★	Slow developing painful nodules, abscesses and ulcers of lower limbs, sometimes extending higher	6
urticaria				★	Acute allergic reaction; sudden plaque like swellings on any part of the body; usually disappear in a few hours	This vol.
worm nodule disease				★	Spread by flies and midges; microfilarial larvae cause small benign nodules under skin in neck and lower limbs	126

*Irritation; nodules; scab formation; oedema under the skin

3.2 Scabs

Scabs, like lumps and nodules, may not be visible and can then only be detected by palpation or parting the wool or hair. If an animal has scab lesions in the skin, it is important to determine whether they cause irritation or not as this can aid diagnosis. Examples of diseases characterised by scab lesions are sarcoptic mange (Fig 2.1) and dermatophilosis (Figs 1.12, 1.13 and 1.14, Volume 2).

3.3 Oedema

Accumulation of fluid in tissues is defined as oedema. Oedema can arise from a variety of causes and when it occurs in the skin, it causes puffy swellings. If poked with the end of the finger, the oedema 'pits on pressure', i.e. the resultant depression in the swelling takes a few seconds to disappear. A well known example is 'bottle jaw' due to sub-mandibular oedema seen in liver fluke and other diseases (Fig 5.2, Volume 2).

3.4 Urticaria and sweet itch

These are diseases of the skin caused by allergic reactions and so can occur anywhere in the world. Urticaria can sporadically affect individual domestic animals but is most common in the horse. Various agents (allergens) can cause it. Oedematous plaques of varying sizes develop very rapidly within minutes or hours of exposure to the causative allergen and can occur anywhere on the body. The condition can appear alarming and, in severe cases, the affected animal may be very distressed, but the clinical signs usually disappear as quickly as they appeared without treatment.

   Sweet itch is a specific allergic skin reaction in horses caused by exposure to *Culicoides* midges. There is inflammation and intense irritation of the skin of the mane, tail and belly. It usually occurs during the warmest months when midges are active, and susceptibility increases with age. In any group of horses, only a few are liable to be susceptible and be affected.

4 Discharges from the mucosal orifices of the head

Some important diseases result in discharges from the mucosal orifices of the head (eyes, nostrils and mouth). It is important to determine whether discharges are confined to one type of orifice (e.g. from the eyes only) or whether they involve more than one type. This indicates whether the problem is localised (e.g. an eye infection) or whether it is a generalised disease that affects several systems (e.g. respiratory and digestive systems). Discharges from the eyes or nostrils may be unilateral or bilateral, which again indicates the degree of localisation of the problem.

As mentioned in Chapter 4, the mucous membranes provide a defence barrier to invading foreign organisms which are trapped by mucus secreted by the membranes. Discharges from the visible mucous membranes indicate that they are responding to an attack, usually by microorganisms. Discharges may be clear coloured (serous) or mucoid due to a significant increase in the secretion of the slimy mucus. Severe insults to the mucous membranes may cause them to become septic and produce yellow/greenish pus resulting in purulent discharges. If mixed with mucus, the discharges are described as mucopurulent.

Discharges from the eyes and nostrils are usually obvious. Those from the mouth, however, are liable to be mixed up with an increase in saliva secretions and so any increase in salivation should always be regarded as suspicious (Fig 1.1).

5   Signs of respiratory disease

Discharges from the nostrils may indicate an underlying respiratory disease involving the lungs as well as the upper respiratory tract. Hence it is prudent to consider respiratory clinical signs along with nasal discharges, and these categories of clinical signs have been included together in some of the following diagnostic tables.

Probably the most obvious clinical sign of respiratory disease is coughing. Coughing is caused by irritation of the lining of the windpipe (trachea) and lower airways (bronchi) that branch from the trachea into the lungs. It is an involuntary reflex action to remove foreign material in these airways, and persistent coughing is an indication that they are diseased. If in doubt that an animal has been coughing excessively, a simple test can be applied. The upper trachea can be gently squeezed. In a normal animal this will produce no response, but if suffering from inflammation or irritation of the trachea, it will cause the animal to cough.

Diseases lower down the respiratory tract in the lungs and lining of the chest cavity cause difficulty in breathing, called *dyspnoea*. The gentle and barely discernible 'three phased rhythm' described in the previous chapter is disturbed and changes to a visible gasping type of breathing, with little or no pause after breathing out. Depending on the severity, the breathing may become harsh and audible; the pain of breathing may cause grunting noises. In an attempt to help its respiration, affected animals may adopt other actions, such as dilation of the nostrils, mouth

Fig 5.4 Contagious bovine pleuropneumonia; note characteristic outward flexion of the elbows and extended head to ease respiration (CTVM).

breathing, stretching out the head and neck, and turning the elbows out (Fig 5.4). The respiration rate may be increased, but this on its own does not necessarily signify lung disease. Animals with a fever are liable to breathe faster and, as pointed out in the previous chapter, healthy animals subjected to exercise or high ambient temperatures also have increased respiration rates.

Diseases of the lungs include inflammation of the lung tissue (pneumonia). The surface of the lungs and the lining of the chest wall are lined by a thin membrane called the pleura. If pneumonia extends to include the pleura, it is called pleuropneumonia which, if extensive, is serious and painful (see Fig 1.2, Volume 2).

NB Lung disease may not necessarily cause the clinical signs above. Animals often develop pneumonia that goes unnoticed except at the abattoir or slaughter point. This is because in general terms the lungs have more than enough capacity for most purposes and can tolerate mild levels of pneumonia providing they are not stressed or overworked.

Diseases characterised by discharges from the eyes, nose or mouth, and respiratory signs are presented in Tables 5.11 to 5.15.

Table 5.11 Diseases of cattle and buffaloes with laboured or rapid breathing

	Cough	Fever	Nasal discharge	Epidemiological picture	Vol. 2 page
anaplasmosis		★		Tick-borne; widespread in tropics; very important in exotic livestock; often fatal	126
bloat				Distension of left flank from trapped rumen gas; laboured breathing in severe cases	190
CBPP*	★	★		Major contagious lung disease; frequently fatal	2
cyanide poisoning				Follows ingestion of young cyanogenetic plants, e.g. sorghum; usually fatal	198
East Coast fever	★	★		Major disease in E. and C. Africa; often fatal; oedema in lungs causes moist cough	131
heartwater		★		Tick-borne disease; acute cases often have oedema in lungs	129
lungworm	★			Chronic lung disease of grazing livestock; rare in tropics	147
malignant catarrhal fever		★	★	Sporadic seasonal disease; blocked nostrils from discharges causes mouth breathing	6
nitrate/nitrite poisoning				Follows grazing of heavily fertilised forage; can cause rapid death; mucous membranes blue	198
shipping fever	★	★	★	Follows some stress factor; often fatal if not treated	177
tuberculosis	★			Major chronic debilitating infection; commonly involves lungs	66

*contagious bovine pleuropneumonia

Table 5.12 Diseases of sheep and goats with laboured or rapid breathing

	Cough	Fever	Nasal discharge	Epidemiological picture	Vol. 2 page
bloat				Distension of left flank from trapped rumen gas; laboured breathing in severe cases	190
CCPP*	★	★	★	Major contagious lung disease of goats; 60–90% may die in an outbreak	12
heartwater		★		Tick-borne disease; acute cases often have oedema in lungs	129
lungworm	★			Common in young sheep in sub-tropical areas with winter rainfall; can cause pneumonia	147
pasteurellosis	★	★	★	Respiratory disease of stressed sheep; acute with rapid death in young; pneumonia in older	177
tuberculosis	★		★	Occasional infection of goats; lung involvement causes chronic cough and nasal discharges	66

*contagious caprine pleuropneumonia

Table 5.13 Diseases of cattle and buffaloes characterised by discharges from eyes, ears, nose or mouth

	Discharges from				Epidemiological picture	Vol. 2 page
	Eyes	Ears	Nose	Mouth		
besnoitiosis	★		★		Sporadic skin disease; early cases resemble malignant catarrhal fever with discharges and enlarged lymph nodes	119
earworm		★			Parasitic otitis of cattle in East Africa; infection from contaminated dip tanks	151
eyeworm	★				Usually harmless infection of eyes spread by flies; occasionally causes eye inflammation and ulceration	119
foot-and-mouth disease				★	Highly infectious; causes vesicles in feet, mouth and other tissues causing salivation and lameness	47
haemorrhagic septicaemia			★	★	Acute killer disease, esp. of buffaloes; fever, nasal discharge and salivation before death	4
IBR*	★		★		Widespread; inflammation of upper respiratory tract; fever; eye/nose discharges; sometimes pneumonia	78
inf. kerato-conjunctivitis	★				Eye infection of ruminants; inflammation and ulceration; most recover; affected eyes occasionally rupture	118
Jembrana disease	★		★	★	Disease of cattle in Indonesia which clinically resembles rinderpest (see below)	138
lumpy skin disease	★		★		Fly-borne nodular skin disease; first clinical signs include fever and discharges from eyes and nose	106

Table *cont'd*

Table 5.13 – *cont'd*

	Discharges from				Epidemiological picture	Vol. 2 page
	Eyes	Ears	Nose	Mouth		
malignant catarrhal fever	★		★	★	Contact with wildebeast or sheep; sporadic, usually fatal; discharges, fever and enlarged lymph nodes	6
mucosal disease			★	★	Sporadic rinderpest-like disease (see below); usually fatal	76
rabies			★	★	Infection usually from bite of rabid animal; severe nervous signs; salivation and bellowing; always fatal	58
Rift Valley fever	★		★	★	Mosquito-borne; calves most susceptible; fever, vomiting, nasal discharges, collapse and death	108
rinderpest			★	★	Contagious, usually fatal; erosions of mucous membranes cause discharges with fetid odour, diarrhoea and fever	60
Schistosoma nasalis			★		'Snoring disease', fluke infection of nasal veins; seen in cattle grazing aquatic snail habitat	154
shipping fever	★		★		Follows stress; usually in young cattle; respiratory infection which may develop to pneumonia and death	177
sweating sickness	★		★	★	Sporadic tick-borne toxicosis of calves; moist skin eczema, fever and discharges; sometimes fatal	139
tuberculosis			★		Major chronic debilitating infection; lung involvement causes chronic cough and nasal discharges	66

*infectious bovine rhinotracheitis

Table 5.14 Diseases of sheep and goats characterised by discharges from eyes, nose or mouth

	Discharges			Epidemiological picture	Vol. 2 page
	Eyes	Nose	Mouth		
blue-tongue	★	★	★	Widespread; severe disease in exotic ruminants; fever, feet lesions may cause lameness	104
contagious agalactia	★			Contagious infection of udder, eyes, joints, and genitalia (males); goats more susceptible	11
eyeworm	★			Usually harmless infection of eyes spread by flies; occasionally causes inflammation and ulceration	119
foot-and-mouth disease			★	Highly infectious; vesicles in mouth and feet cause fever, salivation and lameness	47
inf. kerato-conjunctivitis	★			Eye infection causing inflammation and occasionally ulceration; most recover	118
lungworm				Common in young sheep in sub-tropical areas with winter rainfall; can cause pneumonia	147
Nairobi sheep disease	★	★		Tick-borne disease; high fever, dysentery, abortions, collapse and death	130
nasal fly		★		Very common in sheep, rare in goats; fly larvae in nostrils cause sneezing	85
pasteurellosis	★	★		Respiratory disease of stressed sheep; acute with rapid death in young; pneumonia in older	177

Table *cont'd*

Table 5.14 – *cont'd*

	Discharges			Epidemiological picture	Vol. 2 page
	Eyes	Nose	Mouth		
PPR*	★	★	★	Contagious disease of goats and occasionally sheep; diarrhoea and high death rates	20
Rift Valley fever		★	★	Mosquito-borne; acute disease in young; fever, vomiting, nasal discharge, collapse and death	108
rinderpest	★	★	★	Similar to PPR; mainly cattle problem but virus strains in India pathogenic to sheep and goats	60
Schistosoma nasalis		★		'Snoring disease', fluke infection of nasal veins of animals grazing aquatic snail habitat	154

*peste des petits ruminants

Table 5.15 Diseases of horses and donkeys with discharges from the eyes, nose or mouth and/or respiratory signs

	*Discharges		Resp signs	Epidemiological picture	Vol. 2 page
	Eyes	Nose			
AHS** (Circ. form)			★	Spread by midges; haemorrhages in eyes and oedema over eyes and into throat may make breathing difficult	111
AHS** (Pulm. form)		★	★	Spread by midges; fever, laboured breathing, violent cough, frothing nasal discharge and death in about 4–5 days	111
anthrax			★	High fever and oedematous swellings of throat may cause difficulty in breathing and death in 2–3 days	35
besnoitiosis	★	★		Uncommon; early signs include fever, photophobia, and eye/nose discharges; skin lesions follow	119
equine inf. anaemia	★	★		Fly-borne; fever, haemorrhages in eyes and under tongue; nose discharges may contain blood	113
eyeworm	★			Usually harmless infection of eyes spread by flies; occasionally causes eye inflammation and ulceration	119
glanders		★	★	Respiratory tract infection; fever, cough, ulcers in nostrils, laboured breathing; acute (donkeys) or chronic (horses)	32
parascarosis			★	Affects foals; migrating larvae in liver and lungs; coughing, loss of condition and diarrhoea; associated with poor hygiene	161
Schistosoma nasalis		★	★	'Snoring disease'; fluke infection of nasal veins; seen in horses grazing aquatic snail habitat	154
trypanosomosis	★			Fly-borne; usually chronic anaemia and loss of condition; parasites may invade certain tissues including the eyes	121

*Discharges from eyes; nose. Respiratory signs   **African horse sickness (circulatory and pulmonary forms)

6   Changes to the visible mucous membranes

As mentioned in the previous chapter, the mucous membranes can be seen at various points, i.e. the gums, nostrils, eyes (conjunctivas), rectum and vagina. These visible mucous membranes are valuable indicators of an animal's state of health, and veterinarians automatically inspect them when examining a sick animal. In the healthy animal they have a pink colour and glisten from the mucous secreted on their surfaces. In certain conditions, they change colour as follows:

Pale pink to white   Mucous membranes have a rich blood supply and the pink colour is due to red blood cells circulating through them. A shortage of red blood cells (anaemia) results in the visible mucous membranes becoming pale or even white if the anaemia is severe (Figs 5.5 and 5.6). There are many causes of anaemia. Some important tick-borne diseases attack and destroy red blood cells, other diseases impair the animal's ability to produce new red blood cells in the bone marrow.

Yellow   As just mentioned, anaemia can be caused by destruction of red blood cells. Such destruction leads to release of their reddish pigment (haemoglobin) which is converted to another pigment (bilirubin). If excessive, the bilirubin stains tissues, including the visible mucous membranes, a yellowish colour (jaundice). Thus such diseases may cause the visible mucous membranes to be both pale and yellow due to anaemia and jaundice, such as the tick-borne diseases anaplasmosis, babesiosis and theileriosis. Babesiosis may result in so much release of haemoglobin into the blood stream that some of it is excreted unchanged in the urine, which is red in colour as a result. This is called haemoglobinuria or, more commonly, redwater (Fig 5.7).
  Bilirubin is also a by-product of the liver's complex metabolism, and in healthy animals it is passed in bile from the liver into the gall-bladder and then excreted in the faeces. Anything that impairs the flow of bile, such as liver disease, results in bilirubin accumulating in tissues. A well known example is fasciolosis, where the invading liver flukes cause inflammation (hepatitis) and fibrosis (cirrhosis) in the liver (Fig 5.8). The end result is again jaundice but usually more intense than the jaundice caused by the breakdown of red blood cells.

Red   In contrast to the paleness of anaemia, the visible mucous membranes and other tissues may become congested with blood. As this is a common sequel of so many diseases, however, their tabulation would be of little value and has not been included in this book. For example, any disease that causes a fever is also liable to cause congestion of tissues.

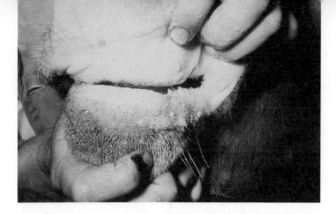

Fig 5.5   Anaemic mucous membranes in mouth of an ox with babesiosis (CTVM).

Fig 5.6   Anaemic conjunctival mucous membranes in sheep with haemonchosis (CTVM).

Fig 5.7   Haemoglobinuria or 'redwater' in bovine babesiosis (CTVM).

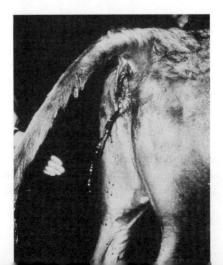

Fig 5.8 Severe hepatitis and cirrhosis of the liver of an ox caused by fasciolosis (CTVM).

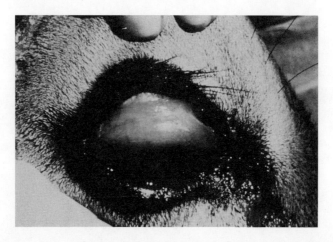

Fig 5.9 Bovine petechial fever with characteristic conjunctival congestion and haemorrhages (CTVM).

Nevertheless one disease which causes reddening of the mucous membranes has been singled out in the diagnostic tables, cyanide poisoning which is caused by ingestion of certain plants (see Chapter 6, Volume 2). Blood carries oxygen from the lungs to the tissues, but cyanide poisoning blocks this process so that the oxygen remains in the blood. The resultant bright red colour of the visible mucous membranes is very distinctive.

Blue In contrast to cyanide poisoning, some conditions result in impaired oxygenation of the blood, causing tissues to become blue in colour, called *cyanosis*. Again, this can be seen in the visible mucous membranes. In animals suffering from nitrate/nitrite poisoning (see

Chapter 6, Volume 2), the haemoglobin in blood is converted to methaemoglobin which is unable to transport oxygen to tissues, resulting in cyanosis.

6.1 Haemorrhages

Haemorrhages from the small blood vessels in mucous membranes are seen as small red 'splashes'; small 'pin-point' haemorrhages are called petechiae, whereas larger ones are called ecchymoses. Petechial or ecchymotic haemorrhages commonly result from a condition called septicaemia, in which infectious organisms and their toxins are circulating in the blood. Septicaemic animals usually have a fever and are usually very ill. As for tissue congestion, because septicaemia and haemorrhages in the visible mucous membranes can be a sequel of so many diseases, no attempt has been made to tabulate them. Petechial fever, however, has been singled out in the diagnostic tables as petechial haemorrhages in the visible mucous membranes is especially characteristic of this disease (Fig 5.9).

7 Fever

An increase in body temperature arises in so many diseases that this clinical sign on its own is of little diagnostic value. However, presence or absence of fever can be helpful in differentiating different diseases with otherwise similar clinical signs and for this reason, this has been included in Tables 5.16 and 5.17 which present diseases with changes in the visible mucous membranes.

8 Diarrhoea

As pointed out in the previous chapter, the presence of diarrhoea may merely reflect a temporary change to animals' normal faeces while adapting to a change in diet. Hence if examining animals with diarrhoea, it is essential to take this possibility into account. Certain chronic diseases are characterised by diarrhoea, so it is important to determine whether diarrhoeic animals have been ill for a long time or not. Diarrhoea in disease is usually due to pathological changes in the intestines, and such changes may include haemorrhage with blood appearing in the faeces (dysentery). If the haemorrhages occur high up in the intestines, the faeces have a brownish-black colour, but if they are lower down, the haemorrhagic blood is passed unchanged in the faeces which is red in colour; the blood may be obvious as clots. These features are included in Tables 5.18 to 5.21.

Table 5.16 Diseases of cattle and buffaloes with changes to colour of visible mucous membranes

	Visible mucous membranes					Fever	Epidemiological picture	Vol. 2 page
	'Pa.	Yell.	Red.	Bl.	Ha.			
anaplasmosis	★	★				★	Major tick-borne disease; main signs fever, anaemia and jaundice; exotics very susceptible	126
babesiosis (redwater)	★	★				★	Major tick-borne disease; sick cattle have haemoglobin in urine (redwater); exotics very susceptible	135
bovine petechial fever					★	★	Tick-borne disease of E. African cattle; fever, small haemorrhages in mucous membranes and death	139
bracken poisoning							Carcinogen in bracken causes tumours in bladder and haemorrhage into urine; may cause anaemia	197
copper deficiency							Grazing ruminants in many areas; signs include anaemia, poor growth, and loss of hair pigment	187
cyanide poisoning			★				Poison from certain fast growing plants; bright red mucous membranes, convulsions and rapid death	198
fasciolosis (liver fluke)	★						Major problem of ruminants grazing near aquatic snail habitat; emaciation, anaemia and diarrhoea	151
nitrate/nitrite poisoning				★			Grazing forage heavily fertilised with nitrates; nitrite in rumen disrupts blood transport of oxygen	198
PGE** (worms)	★						Major problem of grazing ruminants; signs include loss of condition, diarrhoea and anaemia	143
senecoisis		★					*Senecio* weeds contain liver toxins; chronic liver damage, loss of condition and sometimes jaundice	199
theileriosis	★	★				★	Major tick-borne disease; signs include fever, anaemia, jaundice and enlargement of lymph nodes	131
trypanosomosis	★					★	Major fly-borne disease; signs include chronic loss of condition, anaemia and, often, eventual death	121

*pale; yellow; red; blue; haemorrhagic   **parasitic gastroenteritis

Table 5.17 Diseases of sheep and goats with changes to colour of visible mucous membranes

	Visible mucous membranes				Fever	Epidemiological picture	Vol. 2 page
	Pale	Red	Yell.	Blue			
anaplasmosis	★		★		★	Tick-borne blood parasite infection; fever, anaemia and jaundice; exotics susceptible; rare in indigenous	126
babesiosis	★		★		★	Similar to anaplasmosis; sick animals usually have haemoglobin in urine which is red as a result	135
cyanide poisoning		★				Poison from certain fast growing plants; bright red mucous membranes, convulsions and rapid death	198
fasciolosis	★					Major problem of ruminants grazing near aquatic snail habitat; emaciation, anaemia and diarrhoea	151
nitrate poisoning				★		Grazing forage heavily fertilised with nitrates; nitrite in rumen disrupts blood transport of oxygen	198
PGE* (worms)	★					Major problem of grazing ruminants, particularly sheep; loss of condition, diarrhoea and anaemia	143
schistosomosis	★					Sheep grazing near aquatic snail habitat; clinical signs include anaemia and diarrhoea	154
theileriosis	★		★		★	Tick-borne disease; signs include fever, anaemia, jaundice and enlargement of lymph nodes	131
trypanosomosis	★				★	Tsetse fly transmitted; fairly uncommon in sheep and goats; fever at first; later chronic anaemia	121

*parasitic gastroenteritis

Table 5.18 Diseases of cattle and buffaloes characterised by diarrhoea

	Blood in faeces	Disease type		Epidemiological picture	Vol. 2 page
		Acute	Chronic		
arsenic poisoning	★	★		Contamination of feed with arsenical dip; usually fatal but now rare	203
ascariosis			★	Unthriftiness in calves under 6 months; particularly important in buffalo calves	150
calf enterotoxaemia	★	★		Rare; affects young calves; acute cases die in a few hours	54
castor oil seed cake poisoning	★	★		Accidental feeding of castor cake; may be fatal	202
coccidiosis	★	★		Affects calves in crowded conditions; severe cases may have prolonged convalescence	55
copper deficiency			★	Affects grazing ruminants in many areas; poor condition and diarrhoea	188
Dicrocoelium infection			★	Occasional cause of disease similar to fasciolosis	156
fasciolosis			★	Grazing snail infested pastures; emaciation; can be fatal in severe cases	151
heartwater	★	★		Tick-borne; profuse bloody diarrhoea common in clinical cases	129
Jembrana disease		★		Rinderpest-like disease of cattle in Indonesia	139

Johne's disease	★		★	Rare in tropics; affects adults; chronic emaciation and diarrhoea; usually fatal	50
mucosal disease		★		Sporadic rinderpest-like disease in young cattle	76
neonatal diarrhoea		★		Very important in young calves intensively reared; rare in extensive systems	52
paramphistomosis			★	Uncommon; affects calves grazing near aquatic snail habitat	156
PGE* (worms)			★	Important cause of diarrhoea and unthriftiness in grazing cattle	143
rinderpest	★	★		Very contagious; usually fatal; discharges with fetid odour from eyes, nose and mouth	60
rumen acidosis		★	★	Accidental ingestion of large quantity of grain; can be fatal	201
salmonellosis	★	★		Any age can be affected; can be fatal; rare in extensive systems	64
schistosomosis	★		★	Affects livestock grazing near aquatic snail habitat	154

*parasitic gastroenteritis

Table 5.19 Diseases of horses and donkeys characterised by diarrhoea

	Blood in faeces	Disease Acute	Disease Chronic	Epidemiological picture	Vol. 2 page
arsenic poisoning		★		Contamination of feed; abdominal pain, salivation, collapse and death; now rare	203
castor oil seed cake poisoning	★	★		Accidental feeding of castor cake; may be fatal; abdominal pain, sweating and incoordination	202
helminthosis			★	Associated with worm contaminated 'horse-sick land'; wasting, anaemia, coughing and diarrhoea	161
neonatal diarrhoea		★		Usually associated with overcrowding and poor hygiene; affects foals up to 6 weeks old	52
salmonellosis	★	★		Can affect any age; can be fatal; fever, enteritis and dysentery; unlikely in extensive systems	64
senecciosis			★	Liver damage from prolonged grazing of *Senecio* weeds; loss of condition, jaundice and diarrhoea	199

Table 5.20 Diseases of sheep and goats characterised by diarrhoea

	Blood in faeces	Epidemiological picture	Vol. 2 page
coccidiosis	★	Affects lambs/kids in crowded conditions; can be severe but rarely fatal	55
Dicrocoelium infection		Occasional cause of disease similar to fasciolosis (see below)	156
fasciolosis		Grazing snail infested pastures; chronic debilitating disease; often fatal	151
Johne's disease		Rare in tropics; affects adults; chronic usually fatal debilitating disease; foul smelling diarrhoea	50
lamb dysentery	★	Affects young lambs in first few weeks with plenty of milk; abdominal pain and rapid death	167
Nairobi sheep disease	★	Tick-borne disease; high fever, dysentery, abortions, collapse and death	130
neonatal diarrhoea		Acute potentially fatal disease; important in intensive systems; rare in extensive pastoral systems	52
paramphistomosis		Uncommon; affects sheep/goats grazing near aquatic snail habitat	156
PGE* (worms)		Important cause of diarrhoea and unthriftiness in grazing sheep; less important in goats which browse	143
PPR**	★	Contagious disease clinically similar to rinderpest (see below)	20
rinderpest	★	Contagious; usually fatal; eye/nose/mouth discharges; foul smelling diarrhoea with blood and mucous	60
salmonellosis	★	Any age can be affected; can be fatal; rare in extensive systems; usually acute disease	64
schistosomosis	★	Affects animals, particularly sheep, grazing aquatic snail habitat; diarrhoea often with blood and mucus	154

*parasitic gastro enteritis   **peste des petits ruminants

Table 5.21 Diseases of pigs characterised by diarrhoea

	Blood in faeces	Epidemiological picture	Vol. 2 page
castor oil seed cake poisoning	★	Accidental feeding of castor cake; may be fatal; profuse watery diarrhoea, abdominal pain and vomiting	202
coccidiosis	★	Uncommon; affects young piglets in crowded conditions; profuse diarrhoea and occasional vomiting; up to 20% may die in an outbreak	41
enterotoxaemia	★	Uncommon in tropics; affects sucking piglets up to one week old; depression, diarrhoea with blood and death within 24 hours	167
helminthosis	★	Associated with crowded, unhygienic conditions; affects pigs of any age depending on species; other signs include wasting, vomiting, cough	158
neonatal diarrhoea		Usually associated with young piglets in intensive systems; can be fatal	52
rinderpest	★	Contagious; severe diarrhoea, eye/nose discharges, salivation and death about 10 days (usually) later; now confined to southern India	60
salmonellosis	★	Any age can be affected; can be fatal; rare in extensive systems	64

9 Nervous signs

The central nervous system (CNS), or brain and spinal cord, is exceedingly complex but can roughly be divided into two components, the autonomic nervous system and the sensory system. The autonomic nervous system controls the activity of systems of which the animal is unaware, e.g. the tiny muscles that open and close the pupils of the eye to allow in the required amount of light; the muscles that control sphincters in the urinary and digestive systems so that urine and faeces are not being passed all the time. The autonomic system also controls various glands in the body. The sensory system involves muscles that are controlled by the animal, such as muscles for movement, walking, etc. This system is also involved in the senses (taste, sight, smell, hearing and touch) as well as the animal's general mental state. Thus it can be appreciated that diseases of the CNS may result in a large range of clinical signs depending on the severity and location of the lesion(s).

An explanation of the signs caused by diseases of the sensory systems controlled by the brain would be very complex, but the one feature that they all have in common is a departure from normal behaviour; hence an appreciation of the normal behaviour of healthy animals is essential. Thus any abnormal behaviour, especially if it is sustained for a period of time, is suspicious. This can range from a subtle change of the animal's normal expression and demeanour to a severe change such as obvious aggression, change of voice, sleepiness, and abnormal cravings to eat or drink unusual materials.

Of course diseases of the brain can result in other clinical signs. Again they may range from little more than trembling and nibbling movements to obvious signs such as blindness, walking round in circles, abnormal gait (e.g. high stepping), head pressing, abnormal head movements, incoordination of movement and convulsions (Fig 5.10). CNS

Fig 5.10   Convulsions in a goat with heartwater in which the causative rickettsial organism, *Cowdria ruminantium*, accumulated in brain capillary cells (Keith Sumption).

disease may cause animals to go into spasms and sudden nervous stimulations such as a loud bang are liable to set these off, such as in cases of tetanus (Fig 6.1, Volume 2). Extreme cases of CNS disease may result in loss of control of muscles of movement and inability to walk or stand i.e. paralysis. Rabies, a well known and fatal disease, causes severe inflammation of the brain (encephalitis). The clinical signs of furious aggression or paralysis are well known, but the early signs can be very subtle and little more than a slight change in behaviour or facial expression. Diseases in which there is inflammation of both the brain and spinal cord are called encephalomyelitis.

Diseases of the autonomic nervous system may result in similar signs, for example, signs such as uncontrolled salivation and urination would be suspicious. Diagnosis of the different types of nervous signs requires a great deal of skill and no attempt has been made in the diagnostic tables to differentiate them with the exception of paralysis which should be obvious, even to the untrained eye.

10   Abnormal gait

An animal with an injury or disease of the feet or limbs may present abnormalities of gait that could be confused with nervous signs. Indeed it is very difficult to explain in writing the differences between, say, incoordination and lameness. The easiest differentiation is that the abnormal gait due to a disease of the CNS is likely to be uncontrolled and irregular, whereas that due to a leg or limb lesion is controlled as the affected animal alters its posture or movement to relieve the affected foot or limb.

Because of the slight, but potential, confusion that may arise between determining whether abnormal gait or posture is due to a disease of the CNS or due to localised lesions in the feet or limbs, these have been included in Tables 5.22 to 5.25. As for nervous signs, recognition of the different forms of lameness requires skill and experience, and the tables only differentiate between lameness and stiffness.

Table 5.22 Diseases of cattle and buffaloes with nervous signs and/or disturbance of gait

	*Ner. sign.	Para-lysis	Gait Stiff	Gait Lame	Epidemiological picture	Vol. 2 page
babesiosis (*B. bovis*)	★				Major tick-borne disease; most severe in exotic breeds; parasites in brain cause nervous signs	135
botulism		★	★		Poisoning from organic matter; toxin often from chewing bones etc. due to pica of phosphorous deficiency	205
chlor. hyd.** carbon poisoning	★				Accidental ingestion of dip or dip concentration too strong; twitching, trembling and convulsions; can be fatal	203
coenurosis	★				Rare; from dog faeces; tapeworm cysts develop in brain/spinal cord; nervous signs, blindness and death	156
ephemeral fever			★	★	Mosquito and midge-borne; usually mild disease of 3–5 days; shifting lameness; sometimes severe and fatal	105
ergotism				★	Mycotoxicosis; usually from rye or other cereal; necrosis of extremities (feet, lips, tips of ear and tail) causes lameness	202
foot-and-mouth disease				★	Very infectious; vesicles form in mouth and feet and other tissues causing lameness and salivation	47
heartwater	★		★		Major tick-borne disease; mild to severe; signs include fever, circling, high stepping gait, convulsions and death	129
hypomagnesaemia	★				High-yielding dairy cattle on lush grass/bucket-reared calves fed milk; staggering, convulsions and death	180

*nervous signs   **chlorinated hydrocarbon poisoning
Table cont'd

Table 5.22 – cont'd

	*Ner. sign.	Para-lysis	Gait Stiff	Gait Lame	Epidemiological picture	Vol. 2 page
milk fever		★			High-yielding dairy cows at calving; drop in blood Ca causes incoordination, recumbency and often death	180
muscular dystrophy			★	★	Fast growing calves sucking selenium-deficient dams; stand and walk with difficulty; rare in extensive pastoral systems	189
OP*** poisoning	★	★			Accidental ingestion of dip or dip concentration too strong; uncontrolled salivation, urination, etc., paralysis and death	204
phosphorous deficiency		★	★		Phosphorous deficiency widespread in grazing cattle; general signs of ill-health include stiffness and bone fractures	186
rabies	★	★	★	★	Usually from bite of rabid dog or jackal; aggression, salivation, bellowing, paralysis and death within a week	58
rickets			★	★	Calcium or phosphorous deficiency in young causes failure of limb bones to calcify properly; long bones become bowed	188
sodium deficiency	★				Overworked lactating animals in hot weather on sodium deficient pastures; salt craving; urine drinking, soil licking, etc.	188
tick paralysis		★			Sporadic disease in calves; incoordination, ascending paralysis, and possible death due to respiratory failure	139
theileriosis	★	★			Major tick-borne cattle disease; fever and enlarged lymph nodes; occasional nervous form (turning sickness)	131

***organophosphorous poisoning

86

Table 5.23 Diseases of sheep and goats with nervous signs and/or disturbance of gait

	*Ner. signs.	Para-lysis	Gait Stiff	Gait Lame	Epidemiological picture	Vol. 2 page
botulism		★	★		Poisoning from organic matter; paralysis, tongue protrusion, recumbency and death	205
bracken poisoning	★				Ingestion over long period causes blindness; may occur in hilly areas of tropics	198
chl. hydrocarbon** poisoning	★				Accidental ingestion of dip or dip concentration too strong; twitching, trembling and convulsions; can be fatal	203
coenurosis	★				Rare; from dog faeces; tapeworm cysts develop in brain/spinal cord; nervous signs, blindness and death	156
ergotism				★	Mycotoxicosis, usually from rye or other cereal; necrosis of extremities causes lameness	202
erysipelas				★	Uncommon; infection of colostrum deprived lambs causes arthritis	46
foot-and-mouth disease				★	Highly infectious; vesicles in feet and mouth cause salivation and lameness; often very mild in tropics	47
footrot				★	Major infection of feet of sheep and sometimes goats; usually worst in wet conditions; severity can vary	13
heartwater	★		★		Major tick-borne disease; mild to severe; signs include fever, circling, high stepping gait, convulsions and death	129

*Nervous signs   **chlorinated hydrocarbon poisoning
Table *comt'd*

Table 5.23 – cont'd

	*Ner. signs	Para-lysis	Gait Stiff	Gait Lame	Epidemiological picture	Vol. 2 page
hypomagnesaemia	★				Affects suckling ewes on lush grass; sudden staggering, convulsions and rapid death	181
lambing sickness	★	★			Affects ewes in good condition around lambing; hyperexcitability, tremors, recumbency and often death	180
muscular dystrophy			★	★	Fast growing lambs sucking selenium-deficient ewes; stand and walk with difficulty; rare in extensive pastoral systems	189
OP*** poisoning	★	★			Accidental ingestion of dip or dip concentration too strong; uncontrolled salivation, urination, etc., paralysis and death	204
pregnancy toxaemia	★				Affects ewes in late pregnancy; listlessness followed by nervous signs, incoordination, coma and death	181
rabies	★	★			Sporadic; usually from bite of rabid dog or jackal; aggression, salivation, paralysis and death within a week	58
rickets			★	★	Uncommon; Ca or P deficiency in young causes deformity of bones in limbs which are bowed	188
scrapie	★		★		Progressive fatal infection of adult sheep and occasionally goats; nervous signs plus intense skin irritation	23
tetanus		★	★		Contamination of skin cuts/wounds etc.; initial stiffness then rigid spasms of whole body, convulsions and death	170
visna	★	★			Uncommon infection of brain; nervous signs, incoordination, paralysis and death; affects sheep from 2 years of age	16

*** organophosphorous poisoning

Table 3.24 Diseases of pigs with nervous signs and/or disturbance of gait

Disease	*Ner. sign.	Para-lysis	Gait Stiff	Gait Lame	Epidemiological picture	Vol. 2 page
African swine fever (ASF)	★				Major infectious disease; in acute form fever, incoordination and death; also peracute or rarely, chronic	27
chlor. hyd.** poisoning	★				Accidental ingestion of dip chemical; twitching, trembling and convulsions; can be fatal	203
erysipelas		★	★	★	Common; in chronic form hot painful swellings in joints and spine; may cause permanent joint distortion	46
foot-and-mouth disease				★	Very infectious; fever, lameness and salivation from vesicles in feet and less commonly, in mouth and snout	47
hog cholera	★	★			Very infectious; resembles African swine fever; nervous signs of incoordination, circling, trembling and paralysis	28
OP*** poisoning	★	★			Accidental ingestion of dip chemical; uncontrolled salivation, urination, etc., paralysis and death	204
rabies	★	★			Uncommon; from bite of rabid animal; variable nervous signs; paralysis and death in two days of first signs	58
rickets			★	★	Calcium or phosphorous deficiency in young causes malformation of bones and bowing of limbs; rare in scavenging pigs	188
Stephanurus dentatus		★			Kidney worm; larvae migrating in tissues including spinal cord cause wasting and sometimes paralysis	158
Talfan disease	★	★			Sporadic; viral encephalomyelitis; mainly in young; incoordination and occasional paralysis; most recover	30
Teschen disease	★	★			Rare; severe form of Talfan disease; fever, incoordination, paralysis and death in a few days	30

*Nervous signs   **chlorinated hydrocarbon poisoning   ***organophosphorous poisoning

Table 5.25 Diseases of horses and donkeys with nervous signs and/or disturbance of gait

	Nervous signs	Paralysis	Stiffness	Epidemiological picture	Vol. 2 page
amitraz poisoning	★			Dip chemical poisonous to horses; causes sleepiness, incoordination and constipation	204
botulism		★	★	From bacterial toxin in organic matter; paralysis over 1–2 weeks; terminal recumbency and death	205
castor oil seed cake poisoning	★			Accidental ingestion; diarrhoea, abdominal pain, sweating and incoordination; may be fatal	202
chl. hydrocarbon* poisoning	★			Accidental ingestion of dip chemical; twitching trembling and convulsions; can be fatal	203
dourine	★	★		Venereal disease; parasites may invade nervous system, causing incoordination and paralysis	70
EE**	★			Mosquito-borne infections; range from subclinical to severe (fever, nervous signs, collapse and death)	114
OP*** poisoning	★	★		Accidental ingestion of dip chemical; uncontrolled salivation, urination, etc., paralysis and death	204
rabies	★	★		Sporadic; usually from bite of rabid dog or jackal; aggression, whinnying, paralysis and death	58
rickets			★	Rare; calcium or phosphorous deficiency in foals causes malformation and bowing of limbs	188
tetanus	★		★	Contamination of wounds etc;; initial stiffness then rigid spasms of whole body, convulsions and death	170
trypanosomoses	★			Fly-borne infections; usually chronic emaciation; parasites may invade central nervous system	121

*chlorinated hydrocarbon poisoning   **equine encephalomyelitides   ***organophosphorous poisoning

11   Reproduction disorders

In order to appreciate whether there is a disease of the reproduction system, an appreciation of normal patterns of breeding and giving birth in local flocks and herds is essential, especially as some of these diseases do not result in detectable clinical signs in the breeding adults. These diseases can be classified according to their effects on fertility, pregnancy, the foetus, or the genitalia.

11.1   Fertility

Farmers are always concerned if their livestock are infertile and fail to conceive during the breeding season. Infertile females may not come into oestrus, or return to oestrus after breeding. If infertility is suspected, it is important that the farmer has accurate breeding records for the veterinarian. These will indicate whether the females are coming into oestrus or not, and if they do whether their return to oestrus after breeding is at regular or irregular intervals. Breeding records of extensive flocks or herds, however, are rarely kept and so infertility often goes undiagnosed. In reality, true infertility in which animals fail to conceive is quite rare and many so called cases of infertility are due to early death of the embryo which disintegrates and is absorbed into the dam. As a result the dam may come into oestrus again and give the appearance of being infertile.

11.2   Pregnancy

Once a breeding female has become pregnant, diseases of the womb may cause death and expulsion of the foetus, or abortion. As mentioned above, early embryonic death may result in apparent infertility, thereafter foetal death results in abortion. When abortions occur, it is important to determine the stage of pregnancy as this can provide useful diagnostic information. During pregnancy, the developing foetus is contained in a membranous sac called the placenta which provides protection and nourishment. Some diseases cause inflammation of the

Fig 5.11   Placentitis in bovine brucellosis with characteristic 'leathery' thickening of placenta (CTVM).

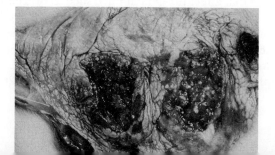

91

placenta (placentitis) which results in abortion, and the appearance of the placenta, if available, may also be diagnostic (Fig 5.11).

Abortions may arise for non-specific reasons, such as high fever. This is a natural defence mechanism to protect the dam, and so it is important to consider this possibility if abortions occur.

Diseases that affect the foetus at the end of pregnancy may cause stillbirths or the birth of live but weak newborn. It is possible to have a range of the different clinical signs (infertility, abortions, stillbirths, weak newborn) resulting in a breeding flock or herd from the same disease, so these are all included in Tables 5.26 to 5.29.

The final effect that may arise from disease or infection of adult females during pregnancy is the birth of live newborn with abnormalities or congenital defects, as in the case of hydrocephalus in Fig 1.6.

Examples – Akabane disease and equine herpes virus Akabane disease is a seasonal viral infection of ruminants transmitted by mosquitoes and midges. Its pathological effects are confined to pregnant females resulting in abortions, stillbirths or the birth of foetuses with severe congenital abnormalities of the brain, vertebral column and joints. Outbreaks have been recorded in Australia, Japan, Korea, Israel and Kenya.

Equine herpes virus infections occur worldwide and one virus, (EHV-1), is an important cause of abortion in horse breeding programmes. Mares infected in late pregnancy are most susceptible and foetuses are usually born dead or die shortly after birth.

11.3 Genitalia

As mentioned, diseases of the genital tract may cause infertility and/or abortion without causing clinical signs in the breeding adults. This does not apply in every case, however, and some diseases also result in pathological changes in the external genitalia that are clinically detectable. Breeding males may have inflammation of the testicles (orchitis). In orchitis, the testicles are hot, painful and may be swollen. If severe, these may be obvious externally, but because of the protective skin of the scrotum, orchitis may only be apparent by palpation. Lesions of the penis may cause discharges from the prepucial orifice. Similarly diseases of the vulva and vagina may also result in discharges.

Mastitis (inflammation of the mammary gland) can arise as a result of certain diseases that affect the genital tract, such as brucellosis in sheep and goats (Table 5.26). The affected mammary glands are hot, painful and may be swollen. Mastitis can, however, occur as a disease entity on its own and indeed is possibly the most important disease of high producing dairy cows (see Volume 2).

All the reproductive diseases are tabulated in Tables 5.26 to 5.29.

Table 5.26 Reproductive diseases of cattle and buffaloes

	'Inf.	Ab.	Stil.	Con. def.	Genitalia lesions		Epidemiological picture	Vol. 2 page
					M.	F.		
Akabane disease		★		★			Seasonal fly-borne disease; abortions and birth of calves with brain and joint defects	This vol.
besnoitiosis					★		Sporadic, chronic skin disease of cattle, commonly with orchitis	119
brucellosis		★	★				Late term abortions in breeding cows	38
campylobacteriosis	★	★					Temporary infertility and occasional abortions	69
epivag					★	★	Rare chronic venereal infection of cattle	72
gangrenous metritis						★	Bruising/tissue damage at calving; usually fatal	167
Iodine deficiency				★			Weak new-born calves with alopecia and goitre	187
IBR**		★		★			Late abortions or calves that die shortly after birth	78
IPV***					★	★	Venereal infection causing inflammation of the genitalia	72
mucosal disease		★		★			Birth of calves with brain and eye defects; newborn blind with abnormal stance and gait	76
Rift Valley fever		★		★			Mosquito-born; pregnant females liable to abort	108
Selenium def.						★	Causes retention of placenta after calving	188
trichomonosis	★					★	Temporary infertility due to early foetal death	73

*Infertility; abortion; stillbirths; congenital defects   **Infectious bovine rhinotracheitis   ***Infectious pustular vulvo-vaginitis

Table 5.27 Reproductive diseases of sheep and goats

	'Inf.	Ab.	St.	WNB	Con. def.	Lesions of genitalia M.	F.	Epidemiological picture	Vol. 2 page
Akabane disease	★	★		★	★			Seasonal fly-borne disease; abortions and birth of lambs/kids with brain and joint defects	This vol.
besnoitiosis	★	★		★		★	★	Goats in Kenya; infertility, abortions, neonatal deaths and cysts on ears/eyes/genitalia	119
Border disease (hairy shakers)	★	★		★	★			Uncommon; congenitally infected lambs have pigmented hairy coats and jerking movements	75
brucellosis		★				★	★	Common; contagious disease causing abortions, mastitis and orchitis	38
chlamydiosis		★	★	★			★	Widespread in sheep but uncommon in extensive pastoral systems; late abortions	9
contagious agalactia		★		★			★	Disease of goats, less commonly sheep; fever, mastitis, arthritis and sometimes abortion	11
copper def. (swayback)					★			Uncontrolled hind-leg movements and paralysis in newborn of Cu deficient dams	187
gangrenous metritis							★	Infection of damaged tissues at lambing; putrid vulva discharges; can be fatal	167

	*Inf.	Ab.	St.	W. N/B.	Con. def.	Epidemiological picture	Page
Iodine deficiency			★	★	★	In endemic areas, high level of stillbirths and weak new-born with alopecia and goitre	187
Nairobi sheep disease		★				Exotic animals susceptible; fever, eye/nose discharges, dysentery, abortions and death	130
Rift Valley fever		★				Mosquito-borne; mild disease in adults but pregnant animals may abort	108
tick-borne fever		★				Fever and occasional abortions; confined to Europe and temperate regions of India	139

*Infertility; abortion; stillbirths; weak newborn; congenital defects

Table 5.28 Reproductive diseases of pigs

	*Ab.	St.	W. N/B.	Con. def.	Epidemiological picture	Vol. 2 page
African swine fever	★				Major disease of pigs transmitted by soft ticks and by direct contact; clinical signs may include abortion	27
brucellosis	★	★	★		*Brucella suis* infection causes chronic disease with abortion, or birth of stillborn or weak piglets; males may have orchitis	38
hog cholera	★		★	★	Very infectious disease similar to African swine fever; infection in pregnancy causes abortion and birth of weak trembling piglets	28
Iodine deficiency			★	★	High incidence of weak new-born piglets with alopecia and goitre	187
JEE**	★	★		★	Mosquito-borne; accidental infection of pregnant pigs causes abortion, stillbirths, or birth of pigs with congenital abnormalities	116

*Abortions; stillbirths; weak newborn; congenital defects   **Japanese equine encephalomyelitis

Table 5.29 Reproductive diseases of horses and donkeys

	Epidemiological picture	Vol. 2 page
besnoitiosis	Uncommon; unknown life cycle; parasites develop in cysts in skin including the scrotum	119
brucellosis	Rare accidental infection, usually from cattle; may cause abortion	38
dourine	Sporadic usually chronic venereal infection; initial oedematous swellings and discharges from the genitalia, and fever; later emaciation and death	70
equine inf. anaemia	Spread by biting flies; fever, eye/nose/mouth discharges and/or haemorrhages; oedema on abdomen, legs and prepuce in stallions	113
Iodine deficiency	Widespread but sporadic; new-born foals from iodine-deficient dams liable to be weak, have lack of hair and goitre	187
equine herpes virus	Contagious disease; widespread and major cause of abortion in horse breeding programmes	This vol.

Table 5.30 Diseases of camels

	Epidemiological picture	Vol. 2 page
anthrax	Major killing disease of livestock; subacute form seen in camels; high fever and death after several days	35
camel pox	Common; worst in young camels; fever, pus filled skin blisters which scab over and form scars in 4–6 weeks; usually on lips; sometimes more extensive	34
eyeworm	Usually harmless infection of eyes spread by flies; occasionally causes eye inflammation and ulceration	119
helminthosis	Can be major problem; anaemia and diarrhoea; critical time late dry – early wet season when herds are concentrated; oedema around head in acute cases	157
Johne's disease	Sporadic disease uncommon in tropics; affects adults; mainly in ruminants but occasionally camels; chronic wasting disease with diarrhoea; usually fatal	50
malnutrition	Affects camels like other livestock; usually from over-grazing after drought etc; emaciation and death; may be complicated by other conditions, e.g. helminthosis	184
mange	Common skin infestation; intense irritation; main sites are neck, inside thighs and flanks; may end in death	101
orf	Raw painful bleeding scabs on head and around mouth; very common in sheep; could be confused with pox	18

Table 5.30 – *cont'd*

	Epidemiological picture	Vol. 2 page
Rift Valley fever	Mosquito-borne infection; fever and deaths in young; abortions in adults	108
ringworm	Uncommon; fungal infection of skin; raised greyish, circular scabs, usually in head and neck; usually associated with overcrowding	63
saddle sores	Lesions on body from badly fitting harnesses, saddles, etc.; raw areas of skin can become infected and painful	176
tick paralysis	Possibly common; usually in young animals; toxins secreted by ticks cause incoordination, ascending paralysis, recumbency and death from respiratory failure	139
trypanosomosis	Fly-borne chronic wasting infections; major problem especially *T. evansi*; signs include fever, anaemia, eye discharges, nervous signs and eventual death	121
tuberculosis	Rare; chronic debilitating infection of the abdominal cavity, organs and lungs; loss of condition and chronic cough	66
worm nodule disease	Small worms spread by flies and midges; microfilarial larvae cause small benign nodules under the skin	126

12   Diseases of camels

The diseases tabulated under various clinical signs in Tables 5.1 to 5.29 do not include those of camels. This is not an oversight, but because documented information of camel diseases is unfortunately much less than that of other animals, it is more convenient to include all those described in Volume 2 in one table, namely Table 5.30.

6 Distribution of diseases

The preceding chapters have outlined the ways different diseases occur and how they may be recognised. Another very important aspect of understanding diseases is an appreciation of where they occur and why. Some diseases have a cosmopolitan distribution and are found worldwide, whereas others have a limited geographical distribution. There are various reasons for this, some due to natural forces and others because of man's interventions.

A knowledge of disease distribution can be invaluable in diagnosis. For example, if confronted with cattle showing clinical signs of anaemia and enlarged lymph nodes in South America, theileriosis can be ruled out (unless the animals have been recently imported) as this tick-borne disease is absent from the Americas. This chapter outlines the distribution of diseases described in Volume 2 or mentioned in this Volume. Cosmopolitan diseases found worldwide are listed in Table 6.1, whereas those with limited geographical distributions are depicted in a series of maps. It must be stressed, however, that these distribution maps are approximate guides and should not be regarded as definitive delineations of the various diseases. Diseases are dynamic events and they can expand and contract their distribution for various reasons, often quite rapidly.

1 Cosmopolitan diseases

1.1 Infectious and contagious diseases

Many important infectious and contagious diseases, including venereal and congenital infections, found in the tropics also occur worldwide. This is because they do not require particular vectors for their transmission, and can pass from animal to animal irrespective of their geographical location. Thus approximately half of the infectious and contagious diseases that are featured in this book are also found worldwide and are shown in Table 6.1.

Table 6.1   Cosmopolitan diseases and parasites of domestic livestock

---

Infectious and contagious diseases

Of livestock in general	*Of ruminants and camels*	*Of pigs*
Anthrax (All)	Caseous lymphadenitis (S,G)	Erysipelas
Brucellosis (All)	Chlamydiosis (S)	Pig pox
Coccidiosis (All)	Dermatophilosis (C,S,G)	
Neonatal diarrhoea (All)	Erysipelas (S)	*Of horses/donkeys*
Rabies (All)	Footrot (S,G)	Dermatophilosis
Ringworm (All)	Johne's disease (C,S,G,Cam)	Equine herpes abortion
Salmonellosis (All)	Malignant catarrhal fever	Ulcerative lymphangitis
Tuberculosis (All)	(C,B)	
	Orf (S,G,Cam)	

Venereal and congenital infections

Campylobacteriosis (C)	Infectious pustular vulvo-	Trichomonosis (C)
Infectious bovine	vaginitis (C,B)	
rhinotracheitis (C,B)	Mucosal disease (C)	

Arthropods and arthropod-borne diseases

Blood sucking and nuisance flies	*Myiasis causing flies*	*Fly-borne diseases*
Black flies (All)	Blowflies (mainly sheep)	Eyeworm (C,S,E,Cam,B)
Mosquitoes (All)	Nasal flies (S,G)	Infectious kerato-
Midges (All)	Warble flies (C,E)	conjunctivitis (C,S,G)
Forest flies (C,E)	(N. hemisphere)	Summer sores (E)
Horn flies and buffalo flies	Lice	Worm nodule
(C,B)	Fleas	disease (C,E,Cam)
Horse flies (All)	Mites	
Sheep ked (S)		
Stable fly (All)		

Helminth infections

Roundworms	*Flukes*	*Zoonotic helminthosis*
Ascariosis in calves (C,B)	Dicrocoeliosis (Rum)	Hydatidosis (Dogs –
Lungworm (Rum)	Fasciolosis (Rum)	man+All)
Parasitic gastroenteritis	Paramphistomosis (Rum)	*Taenia solium* (Pigs –
(Rum,Cam)		man)
	General	*Taenia saginata* (Cattle –
	Horse helminthosis	man)
	Pig helminthosis	*Trichinella spiralis*
	Tapeworms	(Pigs – man)

Diseases associated with husbandry or the environment

Infections	*Plant Poisons*	*Mycotoxicosis*
Clostridial toxaemias (All)	Bracken (C,S)	Aflatoxicosis (P,calves)
Mastitis (C,B)	Nitrate/nitrite (C,S)	Ergotism (All)
Saddle sores (All)	Cyanide (Rum)	
Pasteurellosis (C,S)	Seneciosis (E,C)	

Table 6.1 – *cont'd*

Diseases associated with husbandry or the environment		
Miscellaneous	*Poisons from feed*	*Farm chemicals (dips)*
Bloat (Rum)	Acidosis (Rum)	Arsenic (All)
Heat stress (All)	Cottonseed cake	Chlorinated hydro-
Metabolic disturbances	(gossypol) (P)	carbons (All)
(Rum)	Castor oil seed cake (All)	Organophosphorous
Nutritional deficiencies/		(All)
imbalances (All)		Amitraz (E)
Botulism (All)		

All=all livestock; C=cattle; B=buffaloes; S=sheep; G=goats; P=pigs; E=equines;
Cam=camels; Rum=ruminants

1.2 Arthropods and arthropod-borne disease

Flies Most of the flies of veterinary importance are found worldwide. They vary widely in their abundance, however, and are generally more important in tropical and sub-tropical climates than in temperate climates; consequently fly-borne diseases are more important in tropical and sub-tropical climates and relatively few are found worldwide. For example, horse flies are found worldwide but only in tropical and sub-tropical climates are they present in sufficient numbers to transmit *Trypanosoma evansi*, an important blood parasite of domestic livestock.

Lice, fleas and mites These are truly cosmopolitan skin parasites, and their presence is usually associated with predisposing factors such as poor hygiene and overcrowding.

1.3 Helminth infections

Most of the important helminth infections of livestock are found world-wide although they are heavily influenced by factors such as climate and grazing intensity.

1.4 Diseases associated with environmental or husbandry factors

Most of these occur worldwide. The main exception is poisoning by plants, many of which have very restricted geographical distributions.

2 Diseases with a limited geographical distribution

These are shown in the maps in Figs 6.1 to 6.14.

2.1 Infectious and contagious diseases

As mentioned above, many of the infectious and contagious diseases found in the tropics have a cosmopolitan distribution. Those shown in Figs 6.1 to 6.4 have a limited geographical distribution but some used to be much more widespread. For example, foot-and-mouth disease used to be widespread throughout Europe but as a result of campaigns to control this very infectious disease, it has been eradicated from much of that continent. Similarly rinderpest used to be a major killing plague of cattle throughout much of Asia, Africa and Europe but programmes of control through vaccination and sanitary measures have almost resulted in the eradication of this scourge. It is important to appreciate this because potentially these diseases could recur in countries where they previously occurred if legislative procedures such as statutory controls of livestock movement were compromised in any way.

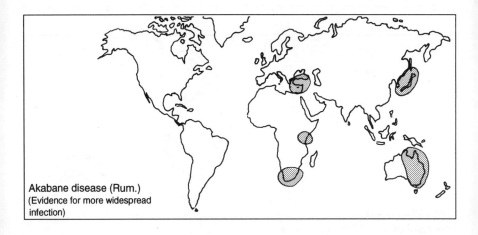

Akabane disease (Rum.)
(Evidence for more widespread infection)

Fig 6.1   Infectious and contagious diseases of ruminants.

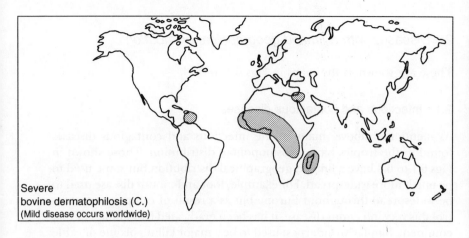

Severe
bovine dermatophilosis (C.)
(Mild disease occurs worldwide)

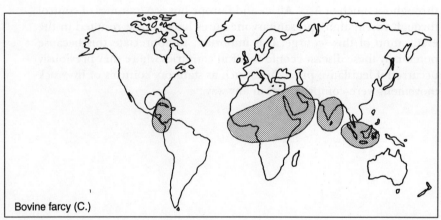

Bovine farcy (C.)

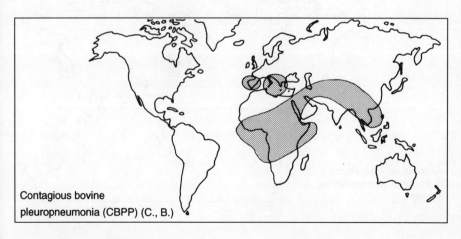

Contagious bovine
pleuropneumonia (CBPP) (C., B.)

Fig 6.1 *cont'd*

102

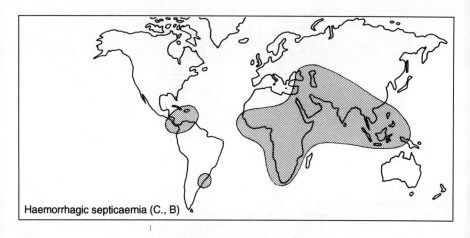

Haemorrhagic septicaemia (C., B)

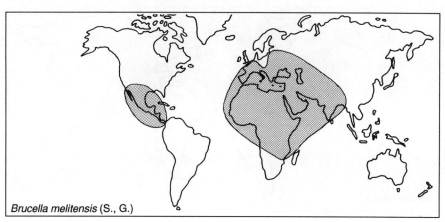

Brucella melitensis (S., G.)

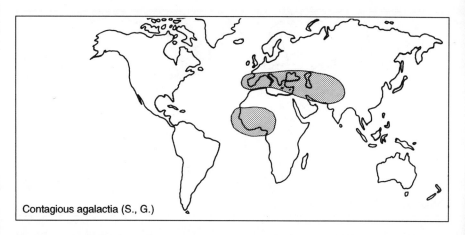

Contagious agalactia (S., G.)

Fig 6.1 *cont'd*

103

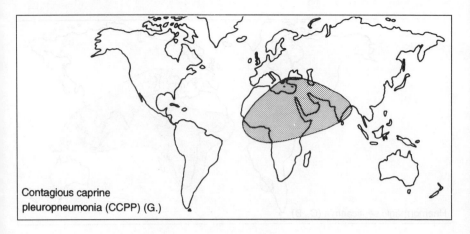

Contagious caprine
pleuropneumonia (CCPP) (G.)

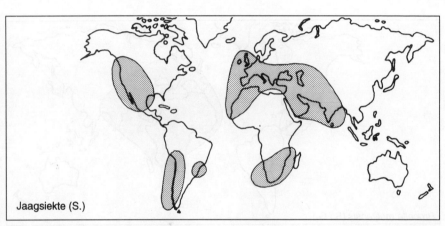

Jaagsiekte (S.)

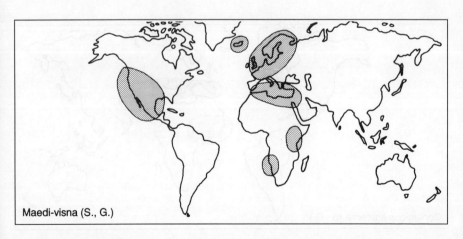

Maedi-visna (S., G.)

Fig 6.1 *cont'd*

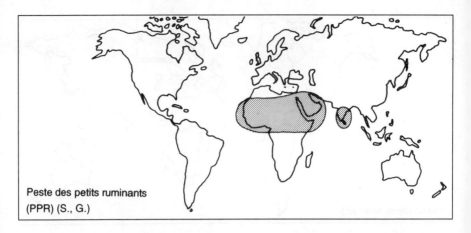

Peste des petits ruminants
(PPR) (S., G.)

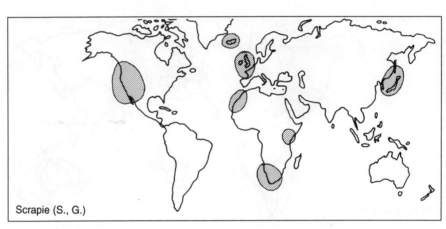

Scrapie (S., G.)

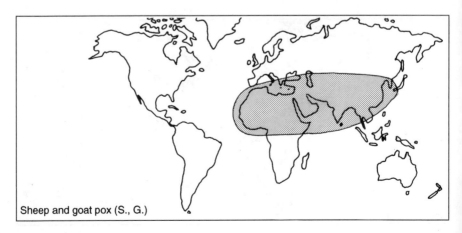

Sheep and goat pox (S., G.)

Fig 6.1 *cont'd*

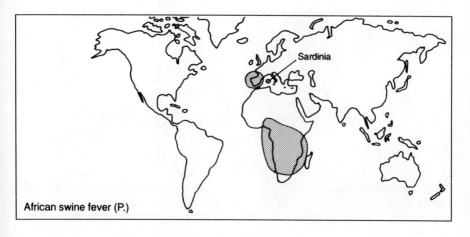

African swine fever (P.)

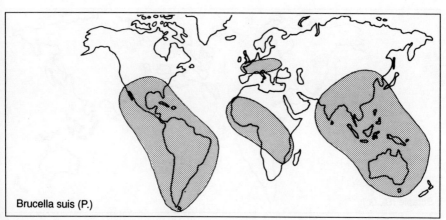

Brucella suis (P.)

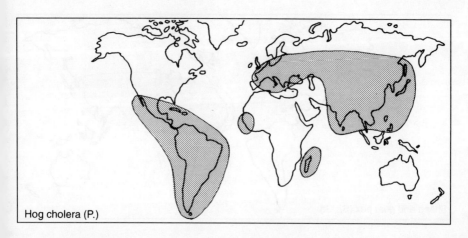

Hog cholera (P.)

Fig 6.2   Infectious and contagious diseases of pigs and horses.

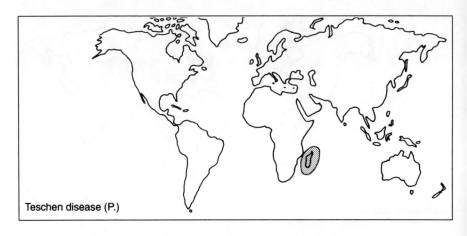

Teschen disease (P.)

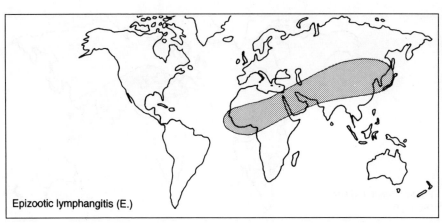

Epizootic lymphangitis (E.)

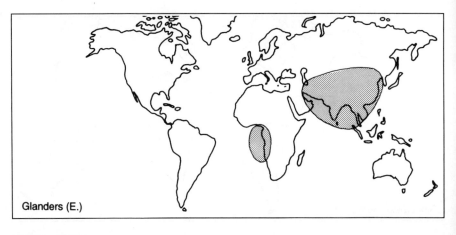

Glanders (E.)

Fig 6.2 *cont'd*

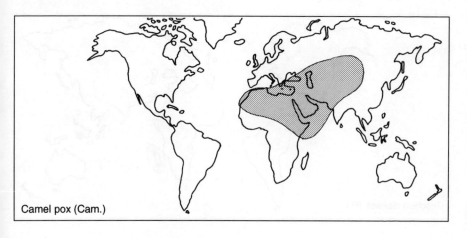

Camel pox (Cam.)

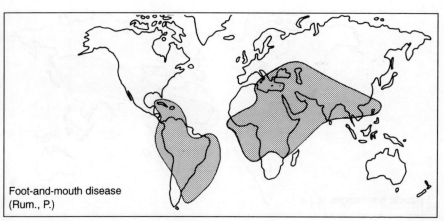

Foot-and-mouth disease
(Rum., P.)

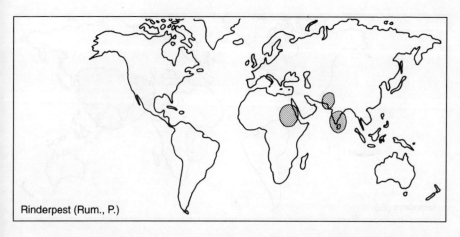

Rinderpest (Rum., P.)

Fig 6.3   Infectious and contagious diseases of livestock in general.

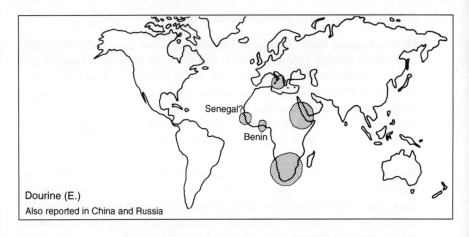

Dourine (E.)
Also reported in China and Russia

Senegal?

Benin

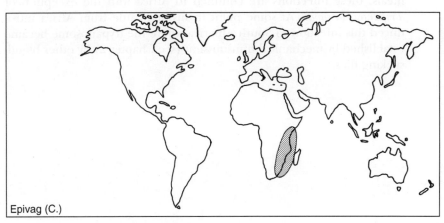

Epivag (C.)

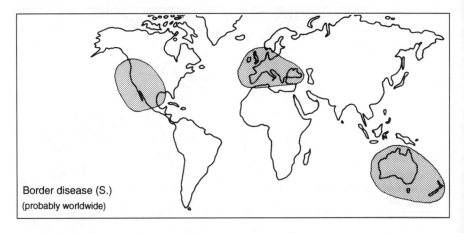

Border disease (S.)
(probably worldwide)

Fig 6.4   Venereal and congenital infections.

2.2 Arthropods and arthropod-borne diseases

Myiasis-causing flies Unlike biting and nuisance flies, some of the flies that cause myiasis (see Chapter 1 and Volume 2) have a limited geographical distribution and are shown in Fig 6.5. Of particular interest is the American screw-worm fly, *Cochliomyia hominivorax* whose 'screw' shaped larvae bury themselves in the tissues of their hosts and cause large foul-smelling lesions. This fly was accidentally introduced into Libya but was successfully eradicated from that country in 1991.

Fly-borne infections These are shown in Figs 6.6 to 6.9. One very important group of infections is the trypanosomoses transmitted by tsetse flies. Tsetse flies are confined to sub-Saharan Africa and as the life cycle of the trypanosome depends on circulation through tsetse fly blood meals, these infections are confined to Africa with the exception of *Trypanosoma vivax*. At some point in history, cattle from Africa introduced this infection to South America where the trypanosome became established by mechanical transmission (see Chapter 2) by other bloodsucking flies.

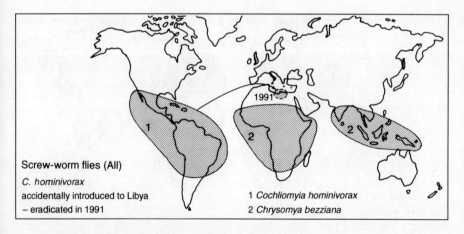

Screw-worm flies (All)

C. hominivorax
accidentally introduced to Libya
– eradicated in 1991

1 *Cochliomyia hominivorax*
2 *Chrysomya bezziana*

Fig 6.5 Arthropods – flies that cause myiases.

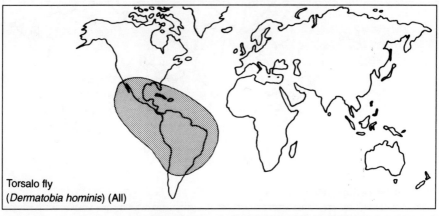

Torsalo fly
(*Dermatobia hominis*) (All)

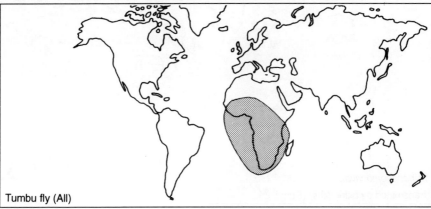

Tumbu fly (All)

Fig 6.5 *cont'd*

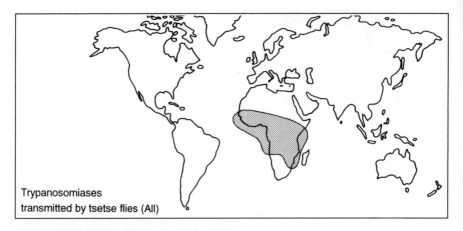

Trypanosomiases
transmitted by tsetse flies (All)

Fig 6.6   Fly-borne trypanosome infections.

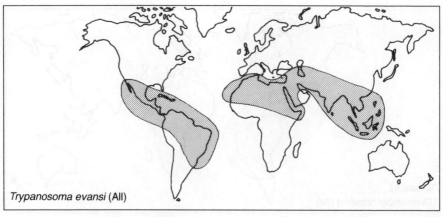

Trypanosoma evansi (All)

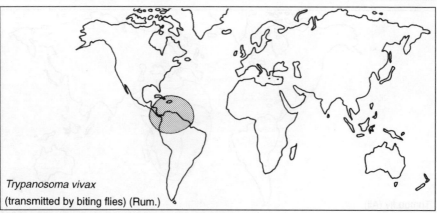

Trypanosoma vivax
(transmitted by biting flies) (Rum.)

Fig 6.6 *cont'd*

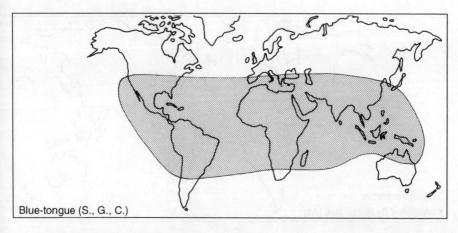

Blue-tongue (S., G., C.)

Fig 6.7   Fly-borne infections of ruminants.

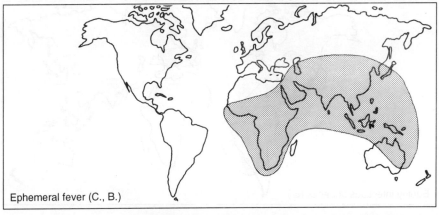

Ephemeral fever (C., B.)

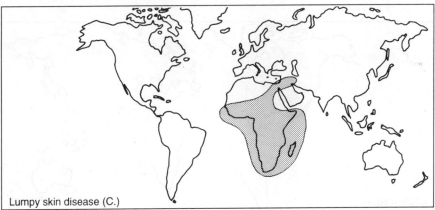

Lumpy skin disease (C.)

Fig 6.7 *cont'd*

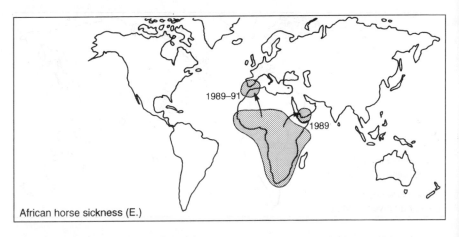

1989–91

1989

African horse sickness (E.)

Fig 6.8   Fly-borne virus infections of horses.

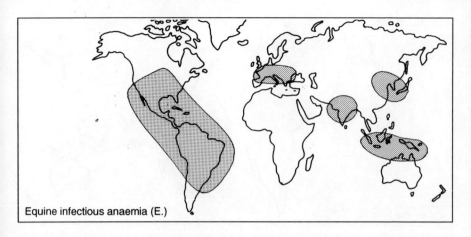

Equine infectious anaemia (E.)

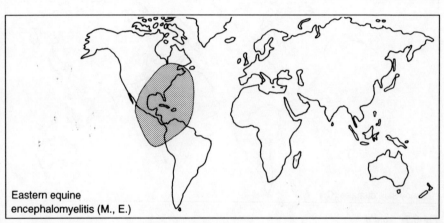

Eastern equine
encephalomyelitis (M., E.)

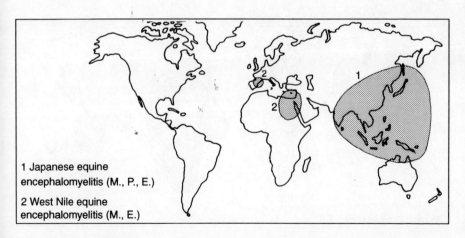

1 Japanese equine
encephalomyelitis (M., P., E.)

2 West Nile equine
encephalomyelitis (M., E.)

Fig 6.8 *cont'd*

114

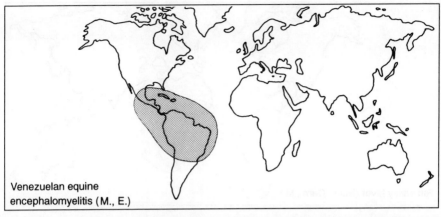

Venezuelan equine encephalomyelitis (M., E.)

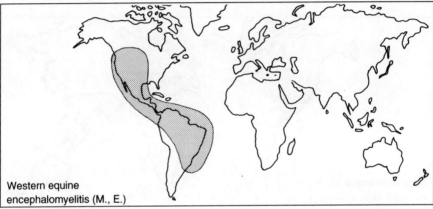

Western equine encephalomyelitis (M., E.)

Fig 6.8 *cont'd*

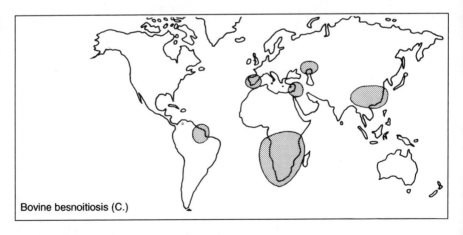

Bovine besnoitiosis (C.)

Fig 6.9   Miscellaneous fly-borne infections.

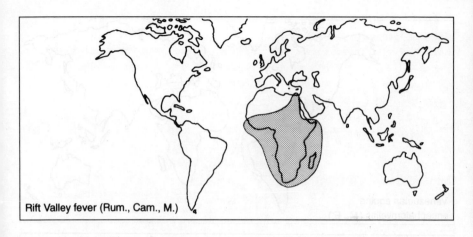

Rift Valley fever (Rum., Cam., M.)

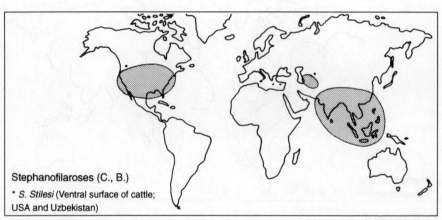

Stephanofilaroses (C., B.)

* *S. Stilesi* (Ventral surface of cattle;
USA and Uzbekistan)

Fig 6.9 *cont'd*

116

Of the fly-borne viral infections, of particular interest is African horse sickness. The virus is transmitted by midges and from time to time wind-borne infected midges introduce the disease into countries neighbouring to sub-Saharan Africa where it is endemic. The resultant epidemics are short-lived, such as recent epidemics in Morocco, southern Spain and the Middle East.

Tick-borne diseases   These are shown in Figs 6.10 to 6.13. Unlike flies, ticks have relatively restricted geographical distributions which largely dictate the occurrence of the diseases they transmit. Some diseases can only be transmitted by ticks and their distribution is thus restricted to that of their vectors, such as the babesioses and theilerioses, heartwater and Nairobi sheep disease. A notable exception is bovine anaplasmosis caused by *Anaplasma marginale*. This rickettsial disease is mainly transmitted by the same species of *Boophilus* ticks which transmit *Babesia bigemina* and *Babesia bovis* to cattle and so it is found in the same parts of the tropics as these two infections (see Figs 6.10 and 6.12). However *A. marginale* can also be transmitted by many other species of ticks, biting flies and by blood-contaminated syringe needles and instruments, etc. Consequently its distribution is wider than that of *B. bovis* and *B. bigemina*.

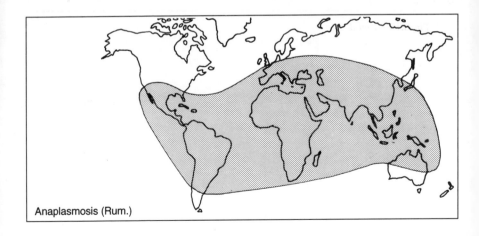

Anaplasmosis (Rum.)

Fig 6.10   Tick-borne rickettsial infections.

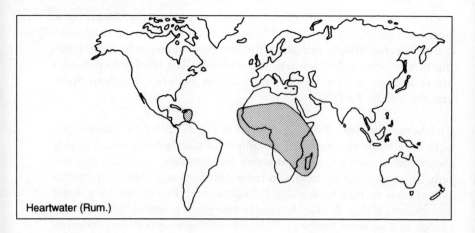

Heartwater (Rum.)

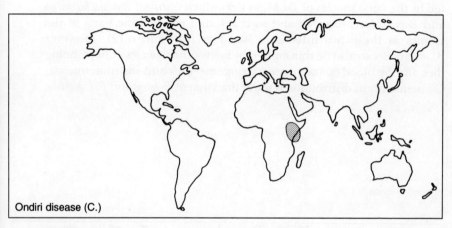

Ondiri disease (C.)

Fig 6.10 *cont'd*

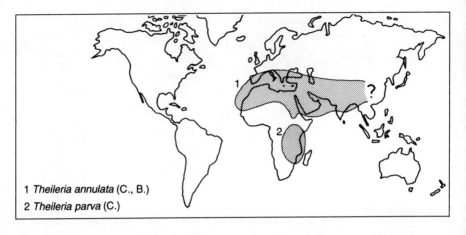

1 *Theileria annulata* (C., B.)
2 *Theileria parva* (C.)

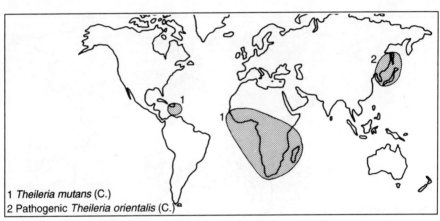

1 *Theileria mutans* (C.)
2 Pathogenic *Theileria orientalis* (C.)

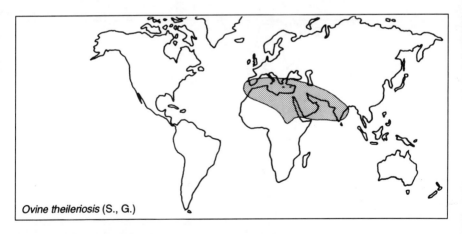

Ovine theileriosis (S., G.)

Fig 6.11   Tick-borne protozoal infections (theilerioses).

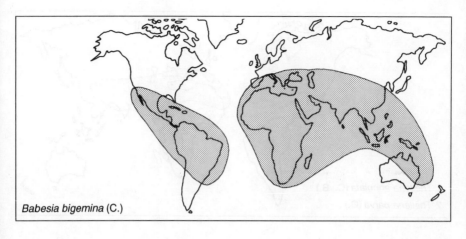

Babesia bigemina (C.)

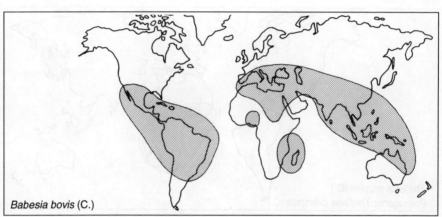

Babesia bovis (C.)

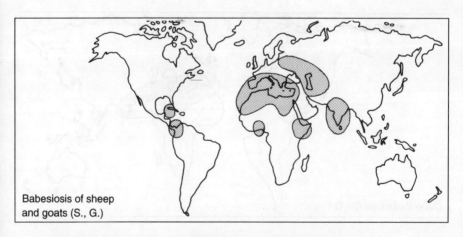

Babesiosis of sheep
and goats (S., G.)

Fig 6.12   Tick-borne protozoal infections (babesioses).

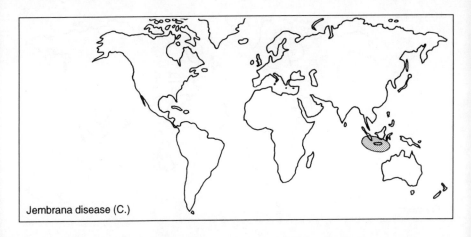

Jembrana disease (C.)

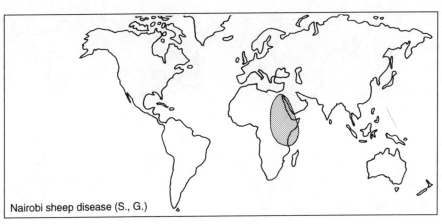

Nairobi sheep disease (S., G.)

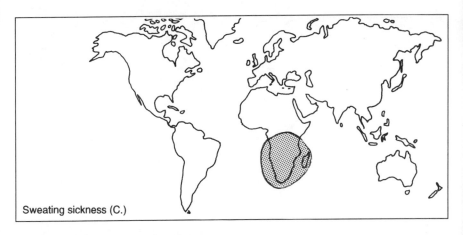

Sweating sickness (C.)

Fig 6.13 Miscellaneous tick-borne diseases.

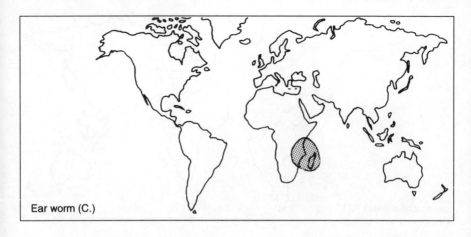

Ear worm (C.)

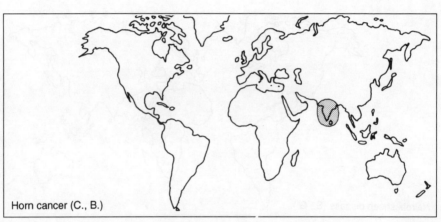

Horn cancer (C., B.)

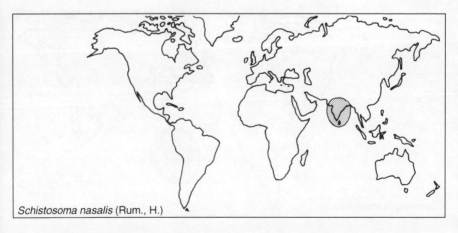

Schistosoma nasalis (Rum., H.)

Fig 6.14   Miscellaneous diseases.

122

7 General veterinary principles

Although it is often beyond the scope of farmers or livestock technicians to take specific action when confronted with disease problems in their animals, nevertheless a number of standard principles and procedures can be applied in most situations. The two most important principles are:

a) healthy animals, well looked after are more likely to cope with any disease agents they confront
b) sick animals are liable to be a major source of disease to healthy in-contact animals, and so should be segregated to reduce the chances of the disease spreading.

1 Health through good husbandry

The best way to ensure the health of animals is to ensure that they are subjected to a good level of husbandry. Healthy animals that are well nourished, provided with water, adequate shelter or shade, and not overworked or stressed are usually better equipped to cope with any disease problems that they may encounter. They are also more productive and it is good economic sense to ensure livestock are in optimal health. Aspects of husbandry are included in books on different animals in this series and are not covered here, but one important point about good husbandry is the knowledge that farmers or herders build up about animals in their care. The good farmer knows his/her animals and is the first to sense if anything is wrong. If consulted because of a disease problem, the first thing that a veterinarian does is to build up a history of the problem by questioning the farmer. This may be essential information and a poor farmer who cannot provide it can actually hinder the veterinarian in applying appropriate action. It must also be said that there are poor veterinarians who do not make the necessary enquiries to the farmer. Farmers and veterinarians are both experts in their respective fields, and the exchange of information between them is an essential component of good animal health and husbandry (Fig 7.1).

2   Preventing disease

As well as good husbandry to ensure the optimal health of animals, it is often possible for farmers to take common-sense measures to prevent or minimise the chances of diseases occurring. The source of many disease agents and parasites are animals themselves, often adults that have acquired them earlier in life and have developed an immunity to them. Thus younger animals, when in contact with older or adult animals, are inevitably exposed to a whole range of potential disease agents. So can anything be done to prevent disease if infection is inevitable? The answer is a definite 'yes', and some of the steps that can be taken are outlined below.

2.1   Housing or shelter

Most forms of animal husbandry include the gathering together of animals for various reasons. At one end of the scale, extensively ranged nomadic or pastoral flocks and herds may be gathered in at night for security reasons whereas in intensive systems, the animals may be housed permanently and food brought to them. Whatever the form of husbandry, holding areas are one of the most likely places for disease agents to build up or where susceptible animals are at risk from contact with infected or sick animals. Even without any specific disease knowledge, these risks can be minimised.

Fig 7.1   The herder and the veterinarian – two 'experts' who can learn from each other.

Permanent housing This often poses the greatest risks to animals. It should be spacious enough to allow animals to lie down or move around without excessive crowding. Good ventilation is essential as this prevents ambient temperatures becoming high and possibly endangering animals to heat-stress. Good ventilation also reduces the risk of spread of respiratory infections.

It is very important that permanent housing is kept clean to prevent a build up of faecal contamination which, as well as being a potential source of pathogenic organisms, attracts flies. The surfaces should be smooth, impervious and free of cracks so that they can be washed down and disinfected at regular intervals. Cracks in the walls of animal houses are particularly important because as well as being difficult to clean properly, they can also become colonised by ticks which can be vectors of a range of important disease agents. Feed and water troughs should be sited so that animals cannot defecate in them.

The make-up of animal groups in permanent housing obviously depends on the type of management but, with the exception of dams suckling their young, mixing of animals of different ages should be avoided if possible.

Coccidiosis is a very good example of the risks of mixing animals of different ages in a confined space. Older animals may be clinically normal but excrete the infectious stage of the intestinal parasites in their faeces and young susceptible animals joining them are liable to become infected and develop the disease. It must be stressed, however, that it is neither practicable or sensible to attempt to rear farm animals in a totally disease-free environment. Domestic animals that are housed, whether permanently or temporarily, are inevitably exposed to a multitude of disease agents throughout life. The sensible approach to their management as outlined should, however, minimise the levels of exposure and the chances of serious disease, but not prevent a build up of immunity.

Temporary shelter For many pastoralist groups, the problems of permanent housing never arise. It would thus be tempting to assume that the risks of disease are negligible, but this could be a dangerous assumption. Extensive flocks and herds are usually held in some form of enclosure at night for security, and some of the disease hazards of permanent housing apply to these also. Animals of different age groups in an overcrowded, dirty holding pen may be just as much at risk to disease as their counterparts in a solid concrete house. Again the approach is largely one of common sense. Enclosures should be spacious so that animals can move around and lie down comfortably (Fig 7.2) and be kept clean to avoid a build up of faecal contamination.

A good approach is to change holding areas at regular intervals. The

Fig 7.2   Animal enclosures should provide shade and be spacious enough for animals to move around and lie down comfortably.

benefits of this have been demonstrated in parts of West Africa where, during the rainy season when farmers are occupied in cultivation, N'Dama cattle are tethered for much of the day in holding places. Unless these holding places are changed every three weeks or so, they become heavily contaminated with a build up of infectious worm larvae hatched from worm eggs passed out in the faeces. As a result the cattle in them, particularly weaned calves, acquire significant worm burdens to the detriment of their general condition and their ability to cope with the hardships of the following dry season. This specific example can be used as a corollary for many similar situations.

2.2   General hygiene

The above section has outlined the importance of good hygiene and management of animal accommodation. This approach should be extended to all aspects of animal husbandry, and all utensils, harness equipment, etc. should be kept clean and disinfected. Failure to do so results in equipment encrusted with organic matter which is an ideal breeding ground for harmful bacteria. Dirty equipment also poses the risk of spread of pathogenic organisms by fomites (see Chapter 1).

Disinfectants and antiseptics   Disinfectants are chemicals that either kill or prevent multiplication of pathogenic micro-organisms, and there

126

Table 7.1 Disinfectants and antiseptics

	*Dis.	Antis.	Comments
Oxidising agents			
Hydrogen peroxide		★	Mild action. Used for cleansing skin abscesses
Potassium permanganate		★	Used for wounds. Stains the skin
Halogens			
Sodium hypochlorite		★	Used as a teat dip in dairy cattle mastitis control
Iodine		★	Used for wound dressings. Stains the skin
Iodophors		★	Udder wash and teat dip in mastitis control
Reducing agents			
Formaldehyde	★		Used for fumigation of buildings
Phenols and cresols			
Lysol	★		General use. Toxic to dogs and cats
Chloroxylenols			
Parachlorometaxylenol	★	★	General all round use
Dichlorometaxylenol	★	★	Similar to above but stronger
Washing soda	★		General washing and disinfection
Aluminium sulphate		★	Used for dermatophilosis in sheep

*Disinfectants; Antiseptics

are many available on the market for a variety of uses. Some of these that can be used in the farm situation are shown in Table 7.1. Manufacturers' instructions should be followed carefully, particularly regarding the following:

a) many disinfectants have to be diluted in water and the correct concentration is essential
b) mixing disinfectants should be avoided as this may make some disinfectants ineffective
c) some disinfectants become ineffective if they contain a build up of dirt and organic matter and have to be changed at regular intervals.

If no disinfectant is available, general hygiene and cleanliness is even more important. Harness equipment should be regularly checked to ensure that it fits properly, and does not traumatise the animal when in use. Dirty ill-fitting equipment is asking for trouble.

Antiseptics are similar to disinfectants, but they can be applied direct to the skin for the treatment of wounds, etc. (see below).

Syringes and needles Of special importance are syringes and needles as many farmers now own their own and administer drugs themselves. We are now in the era of disposable syringes and needles, so that a sterile syringe and needle can be used for every treatment. This will ensure that there is no risk of organisms being passed from one animal to another on contaminated needles, but it poses the problem of safe disposal of the old needles and syringes, a procedure that few farmers are in a position to do effectively. The alternative is to use conventional syringes and needles and sterilise them regularly. For most day-to-day situations, a farmer should only be injecting a small number of animals in any given day, and a small stock of clean sterilised syringes and needles always available should suffice to ensure that a sterile syringe and needle is used for each animal. If more than one animal is being injected with the same drug, then it is in order to use the same syringe as long as a sterile needle is used for every animal. At the end of each day any syringes used should be dismantled and thoroughly cleaned to ensure that they contain no animal tissue or drug residues. Needles should be cleaned by squirting clean water through them. Needles and syringes can then be sterilised by soaking in disinfectant or boiling in water for about 15 minutes.

A major disadvantage of conventional syringes and needles is the risk of transferring pathogenic organisms from one animal to another if they are not properly cleaned and sterilised. Conventional needles also become blunt with repeated use whereas disposable ones are optimally sharp.

Animals In considering general hygiene in animal premises, it is important to include the animals themselves. Healthy animals groom themselves and keep themselves clean, but sick ones may not have the energy or inclination, and have to be kept clean by the owner. Failure to do so may result in sick animals being infested with ectoparasites such as fleas and lice. Skin soiled with faeces may attract flies and expose them to the risk of myiasis (see Chapter 1 and Volume 2).

A particularly important aspect of animal hygiene is care of any wounds that arise. The skin is one of the most important barriers against harmful organisms, and neglected skin wounds, cuts or abrasions can have serious consequences. Again common sense should be the order of the day. Small cuts etc. should heal themselves provided they are kept clean. If necessary, an antiseptic (see Table 7.1) can be applied to the skin to prevent secondary infection with bacteria. Should a wound become septic and infected, however, antiseptics may cause more harm than good and an antibiotic preparation should be used (see below).

2.3 Animals at pasture

Grazing animals scattered on abundant pasture are probably in the safest possible environment with respect to disease risk. They are under no stress and the danger of infectious diseases spreading is minimal. The rainy season may be the safest of all, because grazing is liable to be most abundant, with livestock scattered over wide areas. Rainy seasons are not without their dangers, however, as arthropods are likely to be active and the risks of fly or tick-borne diseases may be greatest during the rainy season. Farmers thus need to acquire the necessary local knowledge about arthropod-borne diseases in their area and take what steps they can. For example, camel herders in Somalia understand that the most serious camel disease, *Trypanosoma evansi* infection, is spread by biting flies and the worst time is during the rainy season when flies are abundant. Thus by the simple expedient of herding sick camels separately, the chances of biting flies spreading the infection to healthy camels are greatly reduced. Greater details of the control of arthropod-borne diseases are provided in Volume 2.

Fig 7.3 As the dry season progresses, extensively reared livestock concentrate on grazing around permanent water supplies. This can provide suitable conditions for the spread of worms, coccidiosis and other diseases.

Poisonous plants   Animals at pasture may be at danger from poisonous plants, and there are two main situations to be considered.

a) Animals introduced to unfamiliar pastures   A phenomenon of grazing livestock is that they learn which plants are dangerous and avoid them if at all possible. Newly-introduced animals may be at risk, however, as they have not acquired the essential 'local knowledge'. Farmers should be aware of this, and try to restrict the grazing time of new animals each day until they have become familiar with the new pasture.

b) Grazing shortage   When pastures are overgrazed or scarce due to a drought, some poisonous plants have deep roots and may be the only vegetation that is green. Hungry cattle may thus overcome their aversion for the plants and eat them. The solution is to provide fodder during such times, but of course this is often sadly beyond the means of farmers.

Grazing shortages arise for a variety of reasons, sometimes due to overstocking as a result of poor management, and at other times due to drought. In general terms, providing seasonal rains do not fail, grazing animals put on condition over the rainy seasons when grazing is abundant, and this helps to sustain them through the following dry season until the next rainy season. Pastoral communities with extensive flocks and herds gradually retreat from scattered wet season grazing as the dry season progresses, and may eventually be forced to congregate round dwindling water supplies if the dry season is prolonged and the rains fail (Fig 7.3). The net result may be a concentration of livestock in poor condition and increased risk of disease spread, such as of worms and coccidiosis, now recognised as a significant problem of livestock under these conditions. Thus water and fodder should be provided to livestock at several sites under such situations if at all possible because, as well as providing essential sustenance, such a practice may help to reduce these livestock concentrations.

2.4   Vaccinations

In many situations, certain infectious diseases are endemic and pose such a risk to local livestock, that the most practical approach is to vaccinate and hopefully prevent the disease. Vaccines come in various forms, but they are all based on the same principle which is to immunise the animal(s) at risk. In order to appreciate this principle, it is essential to explain a little about the processes of immunisation.

Non-specific immunity   Animals possess a very complex system of cells whose function is to recognise invading substances, including harmful micro-organisms, and take action against them. The first line of defence is a series of cells called leucocytes or white blood cells that are gently circulated around the tissues in the blood stream. These are larger and fewer in number than the oxygen-carrying erythrocytes or red blood cells and two types of leucocytes, neutrophils and monocytes, can migrate from the blood stream into tissues and engulf invading micro-organisms by a process called phagocytosis.

Neutrophils are relatively short lived and their function is to phago-cytose harmful bacteria. Monocytes when they migrate into the tissues become larger in size and are called macrophages. Macrophages live longer than neutrophils, are found throughout tissues and their main function is to phagocytose viruses, bacteria and protozoa that can invade animal cells (see Chapter 3). Neutrophils and macrophages between them are often capable of dealing with invading micro-organisms without further help, but if they are overwhelmed then the animal calls on the next line of defence, another group of leucocytes called lymphocytes.

Specific immunity and lymphocytes   Animals possess a very complex system of immune cells, called lymphocytes, whose function is also to recognise any invading foreign substances such as micro-organisms and parasites. A large pool of lymphocytes continually circulates round the tissues via the blood and the lymphatic system, which mainly comprises a network of lymph nodes connected by lymph vessels (see Chapter 5). There are two kinds of lymphocytes, T and B, named thus because they derive from the thymus and the bone marrow respectively.

The T lymphocytes are remarkable cells with the power to both react directly against and kill the invading organism, and to stimulate macrophages to do likewise. T cells also mediate in the function of B cells which, when confronted with a foreign organism, divide rapidly to form larger cells called plasma cells which secrete antibodies into the blood or other body fluids. Antibodies are complex protein mole-cules called immunoglobulins which act specifically against foreign mi-cro-organisms by destroying them or restricting their harmful effects.

If this complex armoury of cells is successful against the invading micro-organism, the animal recovers and, even more importantly, is probably immune against further attacks from the same organism. This arises because a number of T and B cells will have developed a 'memory' against the organism so that in the event of another 'invasion', they can react immediately and kill off the 'invader'. This acquired ability to react promptly is called immunity, and there are two types; that based on a T cell response is called *cellular immunity*, whereas that based on a B cell response and the production of antibodies is called humoral immunity.

Vaccines   Vaccines are based on the pathogenic organisms but without the harmful effect. This is achieved by altering the organism in some way so that if inoculated into the animal, it stimulates either a humoral or cellular immunity, but does not produce the disease effect of that organism. The different types of vaccines are as follows.

a) Inactivated vaccines   These are based on organisms which have been killed by chemicals, heat or radiation. Inactivated vaccines are generally the safest but their main disadvantage is that they are liable to stimulate a relatively weak immune response and may have to be repeated at regular intervals to produce a useful immunity, i.e. as boosters.

b) Attenuated vaccines   These are based on live organisms which have been altered so that if inoculated into an animal they produce an immune response but cause no disease or, at most, only a mild form of the disease. Attenuation can be achieved by various means, for example, by growing the organism in a laboratory culture; by transferring the organism many times (passaging) through a series of animals or laboratory cultures. Attenuated vaccines are generally more effective than killed vaccines, but as they comprise live material, they generally require more careful handling. They often have to be kept at refrigerator temperatures or even deep frozen, necessitating a 'cold chain' from the point of manufacture to the point of vaccination (Fig 7.4).

Fig 7.4   Vaccinating dogs against rabies. The vaccine is kept on ice up to the time of injection, the last 'link' in the 'cold chain'.

It is not always possible to render organisms harmless by attenuation, in which case it may be necessary to follow up vaccination with an appropriate treatment to prevent the disease effects. This infection and treatment approach has been widely used to immunise animals against tick-borne diseases, such as against heartwater (see Volume 2). This method of immunisation obviously requires the greatest care and ideally should only be done under veterinary supervision.

c) Closely related organisms Infection with some non or mildly pathogenic organisms may stimulate an immune response against closely related, but harmful organisms. Such organisms have been used as vaccines, e.g. *Anaplasma centrale* is used as a vaccine against bovine anaplasmosis, a major tick-borne disease caused by *Anaplasma marginale*. Such vaccines have to be used with care, however, because the vaccine organism may be pathogenic to some extent.

d) Toxoids As was outlined in Chapter 3, some pathogenic bacteria produce their harmful effects by secretion of exotoxins. Vaccines, called toxoids, have been developed against some of these exotoxins so that animals are immunised against the toxin rather than the causative organism. Toxoids are the toxins which have been rendered non-toxic, and are usually produced by growing the organism in laboratory culture and treating the toxin produced with a chemical, e.g. formaldehyde. Toxoids are widely used to vaccinate animals against clostridial toxaemias (see Volume 2).

e) Vaccines of the future The parts of organisms, or vaccines made from them, against which an animal mounts immune responses are called antigens. Micro-organisms and parasites usually have many antigens, of which only some stimulate immune responses that are useful and protect the animal against future attacks. The perfect vaccine would not be a whole organism, either attenuated or killed, but would be made up of only the useful protective antigens of the organism. Such a vaccine would confer as good an immune response as the whole organism, but would be safe and have none of the disadvantages of a live vaccine. Recent spectacular advances in molecular biology, immunology and genetic engineering are now promising such vaccines which could revolutionise both animal and human health.

It is essential to appreciate that vaccines vary considerably in their immunising abilities. Some vaccines confer solid immunity that lasts for several years, others may only give partial protection for a limited period. Modern vaccines against rinderpest, based on live attenuated virus grown in laboratory tissue cultures, give a very strong virtually life-long immunity.

By contrast, the immunity from inactivated vaccines against foot-and-mouth disease lasts less than a year, and so where it is used regularly, animals have to be vaccinated once or twice a year.

Finally, it should not be assumed that vaccines obviate the need for good husbandry. Vaccines only stimulate an animal's immune system, and the animal needs to be in good health for this to be working effectively. For instance, it would be futile to expect good immune responses to vaccines from animals in poor condition and harbouring heavy parasite burdens.

2.5 Protection by passive immunity

Section 2.4 has outlined the immunity resulting from recovering from infection, or from vaccines. This is sometimes called active or acquired immunity. The antibodies from a humoral response, however, can also protect other animals. This happens naturally in pregnancy. Newborn animals do not possess a mature immune system, and they receive antibodies from their mother either through the placenta before birth, or from the first milk called colostrum. These maternal antibodies have a limited duration but serve to protect young animals for the first few months of life until their own immune system is developed. Newborn domestic animals derive most of their maternal antibodies from the colostrum so it is vital that they suck their dams within a few hours of birth as the antibody-rich colostrum reverts to normal milk after a day or so.

Acquisition of maternal antibodies for temporary protection is referred to as passive immunity. The principle of passive immunity can also be used to protect animals at other times. By inoculating a susceptible animal with increasing doses of a pathogenic organism or toxin, the animal develops large amounts of antibody in its serum. Serum produced by this approach is called antiserum or hyper-immune serum and it is often produced in rabbits. Inoculation of antiserum confers immediate but temporary immunity of a few weeks against the organism or toxin in question. Antiserum can be used to treat an existing disease or to protect healthy animals exposed to the disease, e.g. hog cholera, clostridial enterotoxaemias and tetanus (see Volume 2). Hyper-immune sera are expensive to produce, however, and they are rarely used today in routine health care of farm animals.

3 Action required when disease occurs

Even with the highest standards of animal management and hygiene, outbreaks of disease occur and again some common principles can be applied to minimise the problem. First of all it is important to determine whether the disease is infectious or not, and the guidelines at the end of Chapter 1 should help in this. If the disease appears to be infectious, then if at all possible the sick animals should be segregated from the healthy, and sharing of feeding troughs, equipment, etc. should be avoided. If there is a possibility that the disease is spread by flies, then they should be segregated as far away as possible. Controlling flies around the sick animals should be considered and possible methods of doing this are described in Volume 2.

3.1 Nursing sick animals

Regardless of whether the disease in question is infectious or not, sick animals are under less stress if isolated quietly away from the rest of the flock or herd. If they are very sick, they will have difficulty in seeking water and shade which should be provided at all times and in easy reach. Appetite may be impaired, but food should also be readily available. Weak animals unable to sit up should be propped up somehow. This is particularly important for ruminants as gas produced in the rumen may not escape in the recumbent animal. The basic hygiene measures should, of course, be applied to isolated sick animals as well.

3.2 Treatment of sick animals

Beyond basic isolation, nursing and hygiene, the next course of action will depend on local circumstances such as whether veterinary help is available, and whether the farmer has had experience of the disease and knows how to apply treatment. If no veterinary help is available, by referring to the principles outlined in Chapter 1, and by examining the animal(s) for clinical signs as described in Chapter 5, it may be possible for the farmer to make a tentative diagnosis and refer to Volume 2 for more specific information.

In reality, many farmers are not in a position to apply specific treatment, in which case the next best approach may be to administer a non-specific drug effective against opportunistic pathogenic micro-organisms. The principle behind this approach is that a sick animal whatever the cause, is more susceptible to infection with pathogenic micro-organisms, particularly bacteria. Administration of an anti-bacterial drug may thus assist the animal against this. There are two kinds of anti-bacterial drugs, the first are synthesized by man and sometimes referred to as chemotherapeutics, and the second are substances derived from living organisms such as bacteria and fungi, called antibiotics. Sulphonamides (e.g. sulphadiazine and sulphadimidine) are examples of chemotherapeutic compounds that have been in use for over 50 years, although they have largely been replaced by antibiotics.

Bacteria and antibiotics   When examining bacteria from cultures in the laboratory, they are commonly smeared on a glass slide, stained with a mixture of dyes called Gram's stain, and examined under the microscope. Some bacteria take up the violet colour of the stain and are called Gram-positive, whereas those that do not are called Gram-negative bacteria. This is a common way of classifying bacteria and it is also relevant to the use of antibiotics. Some antibiotics have a very narrow range of activity and are only effective against either Gram-positive infections (e.g. penicillin) or Gram-negative infections (e.g. streptomycin). Penicillin and streptomycin are sometimes combined to be effective against a broad range of bacterial infections, but this practice has to a certain extent been superseded by antibiotics that are effective against both Gram-positive and Gram-negative infections, called broad-spectrum antibiotics, such as the tetracyclines (oxytetracycline and chlortetracycline).

Not surprisingly, farmers around the world wish to be in possession of broad-spectrum antibiotics, such as oxytetracycline, so that they can administer them themselves whenever their animals become sick. Veterinarians are concerned about this because indiscriminate and uncontrolled administration of antibiotics eventually results in populations of resistant pathogenic bacteria which are difficult to treat. Overuse of antibiotics may also destabilise the normal bacterial flora of the digestive system which may be colonised by atypical organisms such as fungi, with harmful effects. Strictly speaking, antibiotics should only be available by prescription, so that their use is under veterinary control. The reality is, however, that today it is relatively easy to purchase antibiotics around the world and so it is important that farmers and livestock herders appreciate the dangers of indiscriminate use. A sensible compromise would be for farmers to restrict administration of antibiotics to their most sick animals, and concentrate on good hygiene and nursing for the others.

Antibiotics are now produced in various forms, e.g. as injectables, as ointments and aerosol sprays to be applied to skin wounds, etc., as powders in puffers to administer to skin wounds or eye infections, as intra-mammary infusions for mastitis treatment, and in forms that can be administered by mouth.

NB   Chloramphenicol is a very effective broad-spectrum antibiotic but its use in veterinary medicine has been restricted for two reasons. Firstly, it is effective against human typhoid and for this reason its use in livestock should be restricted to minimise the risk of resistance in the bacteria that cause typhoid. Secondly, it is now known to be toxic and impair bone marrow activity in man. Consequently regular use in livestock should be avoided to ensure that there are no residues in animal food products.

Chemoprophylaxis   This is the administration of chemotherapeutics and antibiotics to prevent disease. Prophylaxis, or the prevention of

Table 7.2 Veterinary antibiotics and chemotherapeutics

	Comments and uses
Antibiotics	
Broad-spectrum (used for treatment of a broad range of bacterial infections)	
Ampicillin	General treatment of bacterial infections
Chloramphenicol	Used for human typhoid; veterinary use restricted
Tetracyclines	Also effective against rickettsial infections, and certain protozoa
Limited range	
Griseofulvin	Used for ringworm (fungal infection of the skin)
Penicillin	Gram +ve bacterial infections e.g. anthrax
Streptomycin	Gram –ve bacterial infections
Tylosin	Mycoplasma infections
Avermectins	Effective against broad range of helminths (but not tapeworms and flukes) and arthropods
Chemotherapeutics	
Effective primarily against bacterial and other infections	
Sulphonamides	Coccidiosis and broad range of bacterial infections
Nitrofurazone	Certain bacterial infections and coccidiosis
Trimethoprim and sulphonamides	Broad range of bacterial infections
Effective primarily against protozoal infections	
Amprolium	Coccidiosis
Imidocarb	Babesiosis; also anaplasmosis (a rickettsial infection)
Diminazene aceturate	Trypanosomosis and babesiosis
Quinapyramine sulphate	Trypanosomosis
Homidium chloride/bromide	Trypanosomosis
Mel Cy	Trypanosomosis
Suramin sodium	Trypanosomosis
Buparvaquone	Theileriosis
Halofuginone	Coccidiosis; can be used for theileriosis
Chemoprophylactics	
Isometamidium chloride	Trypanosomosis
Quinapyramine prosalt	Trypanosomosis

disease, can be achieved by repeat administration of a drug over a period of time, and some drugs (chemoprophylactics) have been developed to produce a similar prolonged effect from one dose. Not surprisingly, the prophylactic use of drugs is often popular with farmers but veterinarians are concerned that the practice is overused with a risk of increasingly resistant populations of bacteria. In general, chemoprophylaxis should only be used when absolutely necessary, e.g. to protect healthy animals in contact with sick infected ones.

Chemotherapeutics and antibiotics have also been developed against micro-organisms other than bacteria, and those referred to in Volume 2 are summarised in Table 7.2. There is also a wide range of chemotherapeutics, and even some antibiotics, against helminth infections called anthelmintics, and they are detailed in Volume 2.

8 General veterinary procedures

The preceding chapters have outlined some of the general principles of animal health. Theory on its own, however, is insufficient when it comes to dealing with sick animals, and this concluding chapter describes some of the routine veterinary procedures that have been referred to in Volume 2 and are within the capabilities of most farmers, herders and livestock technicians. These include techniques involved in the administration of medicines, and techniques involved in taking samples for laboratory tests.

1 *Administration of medicines*

Veterinary medicines can be administered by a variety of routes, i.e. by injection, dosing by mouth including by stomach tube, by topical application, and by intra-mammary infusion.

1.1 Injections

This is by far the commonest and, in many respects, the most convenient method of administering veterinary drugs. For this, syringes and needles are required and, today, many farmers and herders have their own. Syringes are available in a variety of sizes, but for practical purposes, two sizes (20 ml and 5 ml capacity) should cover most situations; syringes with a capacity of 20 ml for large injections, e.g. an antibiotic injection to an ox, and 5 ml for low volume injections.

Needles are also available in a range of diameter and length. Needle sizes are given a G number depending on their gauge; the higher the G number the narrower the bore and needles range from 27G (very narrow) to 14G. For routine farm work, 16G and 19G needles should cover most situations. The following is a suggested selection of needle sizes for routine use:

Injection	Animal	Needle length		Needle gauge	
		Inches	mm	G	mm
Intra-venous (I/V)	Large, e.g. cow	1.5–2	40–50	16	1.5
Intra-muscular (I/M)	Large, e.g. cow	1.5	40	16	1.5
Sub-cutaneous (S/C)	Large, e.g. cow	1	25	16	1.5
I/V	Small, e.g. sheep	1.5	40	19	1.1
I/M	Small, e.g. sheep	1–1.5	25–40	19	1.1
S/C	Small, e.g. sheep	1	25	19	1.1

The above is only a rough guide and in reality, if farmers wish to keep their stock as simple as possible, 1" 16G needles can be used for all purposes. Chapter 6 stressed the importance of hygiene, and that syringes and needles should be cleaned and sterilised before use. In addition, unless disposable syringes and needles are used, it is important to check that they are in good working order. Syringe plungers should fit snugly inside the barrel and not allow any injection fluid to escape up the side; needles should be sharp and straight to ensure injections are made with the minimum of trauma. One important point is the fittings. Several are available and it is important that needles and syringes have the same fitting, e.g. 'Luer' fitting is commonly used today.

As well as attending to the care and maintenance of syringes and needles, consideration has to be given to the animal(s) receiving the injection(s). The animal must be firmly held and possibly the most important person is the one restraining the animal. The site of injection should, in theory, be clipped and sterilised by swabbing with a skin antiseptic; alcohol is often used for this purpose. In practice this is rarely done in the field, and providing the injection site is clean and only sterile sharp needles and clean syringes are used, there should be few problems.

Occasionally animals suffering with metabolic disorders (see Volume 2) have to be treated with large I/V infusions of medicine, for example, up to 800 ml calcium borogluconate to cows suffering with milk fever. For this tubes with a special flutter valve are used to deliver the medicine by gravity from a medicine-bottle held above the animal (see below, **Intra-venous infusions**).

Intra-muscular injections This is the commonest type of injection. It is the easiest to perform and allows rapid diffusion from the injection site into the animal's body. Any large muscle mass is suitable, and the ones commonly used are the rump (adult cattle, buffaloes and pigs), the thigh muscle mass at the back of the hind leg (sheep, goats, pigs, camels and young animals in general), the upper neck (pigs), the middle of the neck (horses and donkeys) and the last third of the neck (cattle and small ruminants). Some drugs cause tissue reaction (e.g. long-acting oxytetracycline) and should be injected into less valuable muscles such as the neck muscles. Such drugs may also be painful, and if injected into the rump or leg they may cause temporary lameness.

Having prepared the site for injection, the syringe is filled with the required amount of medicine, the needle detached and held between the first finger and thumb with the fist clenched, the site is then firmly struck two or three times with the side of the fist and at the next 'strike', the fist is turned and the momentum drives the point of the needle through the skin into the muscle (Fig 8.1). The nozzle of the syringe can then be firmly attached to the protruding top end of the needle.

Before delivering the injection, the plunger must be drawn back a little to ensure the needle has not entered a vein If no blood appears then the plunger can be gently but firmly pushed home so injecting all the medicine into the muscle mass. The syringe and needle can then be withdrawn and the injection site rubbed to minimise any swelling and reaction. Veterinary students are commonly instructed to practice this injection into an orange.

Sub-cutaneous injections These are injections given just under the skin but not into the muscle. Drugs injected S/C are more slowly absorbed into the body than I/M injections. Any site where there is loose skin can be used, and in all animals the most common site is under the skin either in front of, or behind the shoulders. A fold of skin is picked up between finger and thumb and, with the needle attached to the filled syringe, the point of the needle is firmly pushed through the skin into the base of the fold and the drug injected. Care must be taken that the needle does not go right through the fold and appear out the other side (Fig 8.2). If done correctly, the needle should be fairly mobile under the skin; if not then start again because the needle may have been inserted into deeper structures. When completed, the needle is withdrawn and the injection site briskly rubbed.

Intra-venous injections With these, drugs are injected directly into the bloodstream and thus act immediately; I/V injections are thus often used in an emergency requiring a quick effect. Proper restraint of the animal is essential and most farmers should leave I/V injections to the veterinarian. The choice of vein is fairly limited and the most commonly used is the jugular vein which runs along a groove (the jugular groove) on the underside of the neck on either side of the windpipe. With the animal well restrained, the head held up and slightly to one side so that

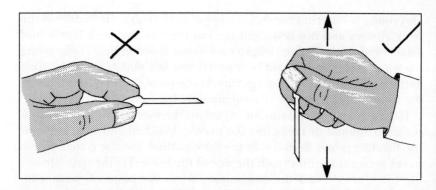

Fig 8.1 For intra-muscular injections, the needle is held in the fist between the finger and thumb.

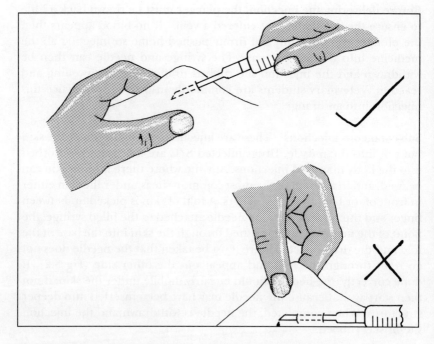

Fig 8.2 Sub-cutaneous injections; care must be taken that the needle is under the skin and does not penetrate through the fold of skin.

the neck is curved, pressure can then be applied with the thumb onto the jugular groove at the base of the neck. As the blood in the jugular vein flows from the head towards the heart, this action causes the vein to fill up with blood due to back-pressure and the distended vein then becomes visible under the skin after about half a minute. If in doubt, the vein can usually be located by gentle palpation; the filled vein has a spongy feel to it. The same effect can be achieved by using a rope round the base of the neck as a tourniquet, and this may be preferred in larger animals. Maintaining the pressure on the vein, the needle can then be inserted into the vein and more-or-less parallel to it so that the length of the needle is lying along the inside of the vein. A common mistake is to insert the needle at too steep an angle to the neck which may result in the needle passing right through the vein (Fig 8.3). It is essential that the needle is really sharp, disposable ones being the best.

If correctly inserted, blood will flow freely from the needle and the filled syringe can then be attached to it. Before completing the injection, the plunger should be drawn back a little to ensure that the needle is still in the vein, as the slightest movement can dislodge it. The pressure on the vein can then be released and the injection carefully, but steadily, completed. If by chance the needle comes out of the vein, this will be detected by a slight change in the pressure on the plunger, and a swelling may appear at the injection site. If this happens, rather than try to relocate the vein and risk causing some degree of tissue damage, it is usually better to withdraw the needle completely and start again at a fresh site. On completion the needle is withdrawn and the site firmly rubbed in the same way as after I/M injections.

Jugular vein 'raised' by applying pressure with thumb

Fig 8.3   Intra-venous injection; the needle should be almost parallel to the raised jugular vein and not at a right angle.

Other veins can be used for I/V injections, but these are best left to the veterinarian. Pigs are particularly difficult, and ear veins or a large vein called the 'Anterior Vena Cava' at the entrance to the chest can be used instead of the jugular vein.

Intra-venous infusions   As mentioned above, occasionally large volumes of medicine in excess of the syringe capacity have to be administered I/V. For these, a medicine bottle attached to a length of tubing about a metre long by a special flutter valve is used. The needle is inserted into the vein as described above, the medicine bottle is then held above the animal until the medicine is flowing freely out of the end of the tube which can then be inserted to the needle by an adaptor. This procedure ensures a steady infusion of the drug by gravity into the vein without introducing any air bubbles (Fig 8.4).

1.2   Medicines administered by mouth

Many drugs can be administered by mouth, but this method of administration is generally more difficult than by injections and animals have to be carefully restrained because there is a risk of the medicine entering the lungs via the windpipe if not carried out properly.

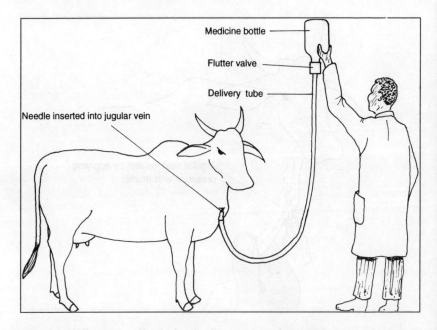

Medicine bottle

Flutter valve

Delivery tube

Needle inserted into jugular vein

Fig 8.4   Intra-venous infusion of large volumes can be administered using a flutter valve to ensure a steady flow from the medicine bottle.

Drenches For liquid medicine given by mouth, the most important aspect is the restraint of the animal in the correct position; its head should be slightly raised and the neck straight so that it can swallow the medicine easily. Drenches can be given using a narrow-necked bottle, preferably strong plastic which cannot break, containing the required amount of medicine. With the animal in the correct position, the neck of the bottle can then be gently but firmly inserted into the side of the mouth, forcing the teeth apart, and tipped up allowing the animal to swallow the contents. The secret to proper drenching is firm restraint of the animal in the correct position, otherwise there is the danger of struggling and either losing some of the medicine or, even worse, the medicine entering the lungs via the windpipe. Drenching guns are also available. These usually have a curved delivery nozzle and although they undoubtedly make the job easier, they have to be used with care as it is all too easy to push the nozzle too far back into the mouth and squirt the medicine into the windpipe by mistake, or even injure the back of the throat.

With the exception of horses and donkeys, all animals can be given liquid medicine by drenching by the above method. Because of the anatomy of the mouth and throat of horses, they can only be drenched by a stomach tube passed down the nostril, and this job should be left to a veterinarian. For this reason, many oral medicines for horses have been formulated to be given in their food or as an oral paste squeezed onto the tongue which the horse then has to swallow.

Tablets, capsules and boluses These are administered by placing at the back of the tongue and then holding the mouth shut until the animal has swallowed it. This requires a little skill, but most farmers and herders can master the art with a little practice. Special 'balling' guns are available if required.

1.3 Topical application

This refers to local application of medicines to the surface of the body, e.g. the skin, eyes, etc. Topical medicines are available as ointments, puffers and aerosol-sprays.

Skin wounds These must first be cleaned by gently applying a piece of clean cloth or cottonwool soaked in antiseptic to remove any dirt or septic discharge material. If no antiseptic or disinfectant is available, boiled water is the next best thing. Any hairs protruding into the wound should be clipped away. This procedure is important as dirty contaminated wounds are liable to heal slowly and attract flies. The form of topical drug application is largely a matter of choice but aerosol-sprays, often containing a dye, are very popular. Aerosols must be used with care round the head to ensure that the eyes are not sprayed accidentally. Wound treatment may have to be given more than once a day until healing is obviously underway. Large wounds may require to be stitched under a local, or even general, anaesthetic and this should only be done by a veterinarian.

Abscesses Infected wounds or skin lesions may develop to abscesses, in which case the lowest point of the abscess should be lanced by a sharp knife or scalpel blade and the pus allowed to escape. The abscess should be left uncovered to drain and general hygiene around the site is important to minimise attraction of flies. Antibiotics may be administered into the abscess 'pocket' to aid healing. Hydrogen peroxide is sometimes inoculated into abscess pockets where its rapid release of oxygen when in contact with organic material provides a mechanical cleansing effect. The lancing of abscesses is not a skilled task, but a common mistake is to make the opening too small to allow all the pus to drain away. For this reason, many farmers may prefer to entrust the task to a veterinarian or a technician.

Eye medicines These are available as ointments in tubes or puffers; ointments should be squeezed directly onto the eyeball under the eyelids and as this can be stressful for the animal, powders puffed onto the eye are often preferred although they are slightly more irritant. Because of the natural washing effect of tears, eye medicines usually have to be repeated several times a day, and the manufacturer's instructions should be followed closely.

One important group of topical medicines are insecticides and acaricides for skin arthropods (see Chapter 2). These are outlined in Volume 2.

1.4 Intra-mammary infusions

Mastitis in cattle and buffaloes may be treated by infusion of specially prepared antibiotic preparations direct into the affected mammary gland(s). The technique is described in Volume 2.

2 Collection of samples for the laboratory

In many cases of sickness or disease outbreaks, a clinical examination on its own may be insufficient to arrive at a diagnosis, in which case laboratory analysis of appropriate samples may be required. The process of collection of samples from clinical cases in the field to their analysis in the laboratory is a complex business involving several disciplines and technical personnel, and sadly this essential part of veterinary science is often neglected. When confronted with outbreaks of disease and with worried farmers looking for immediate action, the easiest option for the veterinarian or technician is often to make a snap diagnosis and act accordingly without attempting to send appropriate samples for laboratory confirmation.

Some sampling procedures are quite complex and should only be carried out by a trained technician or veterinarian. Many of the simpler procedures, however, are well within the scope of farmers and herders and those which have been referred to in Volume 2 are briefly described.

2.1 Blood samples

Laboratory tests on blood samples are amongst the commonest routine laboratory procedures, and ideally blood samples should only be taken by veterinarians or trained technicians. If necessary, however, farmers and herders can take blood samples but this should involve agreeing to some working arrangement with the veterinarian or laboratory staff over the provision of appropriate containers, etc. As everyone knows, following a cut there is bleeding which usually quickly stops because of the blood's ability to clot when it escapes the body. The same thing happens if a blood sample is taken unless a chemical is added to the sample to prevent it clotting. Thus two kinds of blood samples can be taken, clotted samples for serum, or unclotted.

Collection of blood samples   Using the same procedure as described for I/V injections, a sterile syringe needle is inserted into the jugular vein, except for pigs which should be left to a veterinarian. The flowing blood can then be caught in a sterile container such as a test tube and stoppered, or alternatively a clean syringe can be attached to the needle and the blood drawn into it and then transferred to the container. For most purposes a sample of about 10 ml is sufficient. If serum is required for the laboratory, the sample must be allowed to clot which results in clumping together all the red and white blood cells, leaving behind the clear serum. This is done by allowing the sample to stand for about 12 hours in the shade, but not in a refrigerator. Serum should then be separated from the clot, either by carefully pouring it into another container leaving the clot behind, or by removing it using a syringe and needle or pipette. Unseparated samples may be sent to the laboratory, but there is a risk of haemolysis, i.e. leakage of red blood cell pigment into the serum, and haemolysed samples are unsuitable for many tests. Should there be any delay of more than a day or so before sending the samples, again the samples should be allowed to clot and the separated serum can then be placed in the refrigerator until sent. If the delay is prolonged, say several weeks or longer, serum samples can be stored in a deep freeze until required for testing.

Some laboratory tests require whole unclotted blood samples. In this case the sample container must contain an anti-coagulant, a chemical provided by the laboratory to prevent clotting. Several anti-coagulants are available, and the choice depends on the tests to be performed; one is heparin which is the body's natural 'anti-coagulant' to prevent blood clotting in the circulation. Whichever anti-coagulant is used, however, it is essential that the sample is firmly but gently shaken immediately after being taken to disperse the anti-coagulant throughout the sample; failure to do so may result in clotting. Whole blood samples should be dispatched to the laboratory as soon as possible, preferably on ice, because with delay blood cells start to disintegrate. Unless specifically instructed to do so by the laboratory, whole blood samples should not be placed in a deep freeze as this will destroy the blood cells.

Blood sampling has been made very easy with the development of 'Vacutainer®' equipment. Vacutainers® are stoppered vacuum tubes

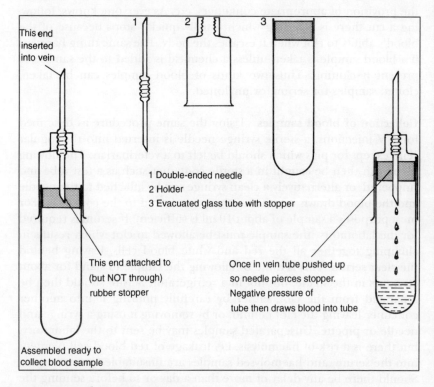

This end inserted into vein

1 Double-ended needle
2 Holder
3 Evacuated glass tube with stopper

This end attached to (but NOT through) rubber stopper

Once in vein tube pushed up so needle pierces stopper. Negative pressure of tube then draws blood into tube

Assembled ready to collect blood sample

Fig 8.5 'Vacutainer'® equipment.

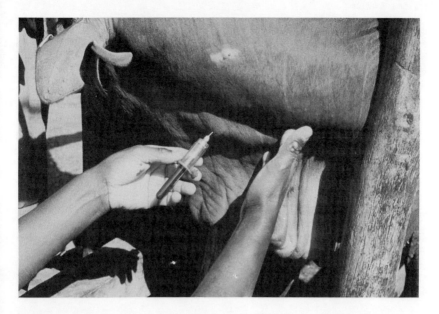

Fig 8.6 Taking a blood sample from the jugular vein of an ox using 'Vacutainer'® equipment (Chris Daborn).

which are available commercially along with special double-ended sterile needles and holders. The needle is screwed into the holder and the point inserted into the jugular vein as described above. The rubber stopper of the tube is then attached to the other end of the needle and the tube pushed to the end of the holder. The negative pressure of the vacuum in the tube then quickly draws the blood into the tube and fills it. Once the blood stops flowing the tube is removed and the needle taken out of the vein (Figs 8.5 and 8.6). Vacutainer® tubes are available in various sizes, with or without anti-coagulant as required. Blood samples collected by this method, as well as being easier to perform, are more hygienic as the 'closed system' ensures the minimum amount of escaped blood. Collecting blood samples by Vacutainers®, however, is relatively expensive.

NB 'Vacutainer®' is the trade name of Becton Dickinson.

Whichever method is used, after the sample has been taken the site should be rubbed and any extraneous blood wiped clean.

2.2 Blood smears

Some very important tick-borne and fly-borne blood infections can be diagnosed from smears of blood on glass microscope slides supplied by the laboratory. Veterinarians working in tropical or sub-tropical coun-

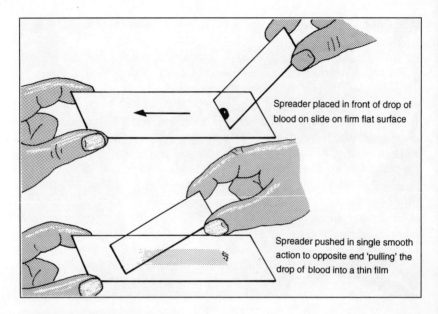

Spreader placed in front of drop of blood on slide on firm flat surface

Spreader pushed in single smooth action to opposite end 'pulling' the drop of blood into a thin film

Fig 8.7   Thin blood film.

tries routinely make such smears from suspect cases of these diseases, and there is no reason why farmers and herders should not do likewise. The best blood for such smears is from small blood vessels, and this can be obtained by pricking the tip of the ear or point of the tail with a sterile sharp point such as a syringe needle. If necessary, the hair is first clipped. This should result in a small drop of blood appearing which is then placed on the end of the slide. Acting quickly before the drop clots, the slide is then held firmly on a flat surface (Land-Rover wings are good for this) and, using the other hand, the end of another slide (the spreader) is placed at an angle of about 30° to 40° just in front of the drop which is allowed to run along the edge of the spreader slide. With a firm but gentle action, the spreader is then pushed to the opposite end, dragging the drop of blood behind it producing a thin blood film. If done correctly all the blood will be spread out ending in a taper (Fig 8.7). Although not essential, the spreader slide should have one corner cut off to ensure the film is narrower than the width of the slide. The blood film should then be dried thoroughly by waving in the air, and placed out of the direct sunlight. Common mistakes to avoid are:

a) too much blood on the slide
b) the spreader is placed behind the drop of blood which is pushed rather than pulled

c) the angle of the spreader slide is too great (results in films being too thick)
d) the film is made in mid-air rather than on a flat firm surface; results in poor control and uneven blood films
e) contamination of the film with dirt or dust.

Dried blood film can be sent to the laboratory for staining and examination under the microscope, if possible after fixation by immersing the slide in methyl alcohol for at least half a minute. The slides should be carefully wrapped in soft paper (e.g. toilet paper) and protected between rigid pieces of cardboard or in special containers, as they will otherwise probably break during transport to the laboratory.

Under certain circumstances when trypanosomosis is suspected, the laboratory may require thick blood films. These again are made by placing a drop of blood on a microscope slide; the drop is then spread with the point of another slide into an area of about 1 cm^2 and thoroughly dried in the air. Thick films, however, must not be first fixed in methyl alcohol as otherwise they cannot be examined in the laboratory.

2.3 Lymph node biopsy smears

Certain diseases result in enlargement of superficial lymph nodes (e.g. theileriosis, see Fig 8.8). Smears of fluid from these can be made for laboratory examination in the same way as for thin blood films. To collect the fluid, a sterile syringe needle of about 16G is inserted into the enlarged lymph node, and a little fluid drawn into it by syringe. The fluid can then be expressed onto a microscope slide and a smear made. The

Fig 8.8 Bovine theileriosis; biopsy smears can be made from the enlarged superficial lymph nodes being demonstrated (CTVM).

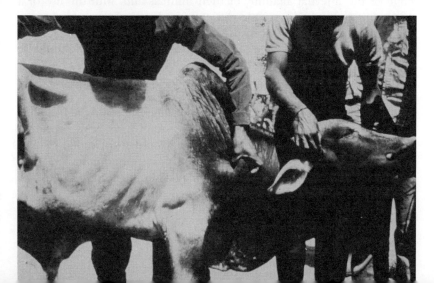

material is liable to contain small pieces of tissue and films are thus somewhat messier than blood films, but this is nothing to worry about.

2.4 Anthrax smears

Special consideration is required for anthrax. When animals die suddenly, or after a short illness, anthrax must always be suspected, in which case the carcass should be immediately burned or buried intact to prevent the risk of the bacillus sporulating and contaminating the environment (see Volume 2). It is also important to check whether such a suspicion is correct or not, and because the disposal of the carcass should not be delayed, it is often up to the farmer or herder to take a sample for the laboratory. For this a blood film is required which can be obtained by nicking the ear of the dead animal, and making a thin blood film as described above, although the film will be considerably messier than the equivalent from a live animal. After taking the sample, the hands should be washed thoroughly because humans are also susceptible to anthrax.

2.5 Autopsy specimens

Autopsy examination of animals that have died is usually a job for the veterinarian who should be equipped with a range of autopsy equipment, containers and preservatives for samples. Occasionally, however, laboratory or veterinary staff may ask the farmer to collect specific samples for laboratory tests, e.g. pieces of lung for the diagnosis of contagious caprine pleuropneumonia; spleen for Nairobi sheep disease; heart blood swabs for haemorrhagic septicaemia. Unless provided with appropriate materials, under such circumstances farmers can only improvise and use common sense. Most farmers acquire a working knowledge of the internal anatomy of their animals and, with the use of a sharp knife, should have little difficulty in locating the organ to be sampled, taking of course every precaution with regard to hygiene in the process. Making the cuts as clean as possible, blocks of tissue should be about 5 cm across. They can then be placed in a dry secure container such as a jam jar with a screw lid and transported to the laboratory as quickly as possible. If there is liable to be any delay, they should be transported on ice in an insulated cool-box. The laboratory may request swabs of certain tissues and, assuming the swabs have been provided, these are handled in the same way. Swabs are usually provided in individual containers which makes security and transportation easier.

For certain laboratory procedures it may be in order to freeze tissues until such time as they can be transported to the laboratory, but this should only be done on the instructions of the laboratory.

2.6   Abortion samples

If abortions occur in a flock or a herd, it is common practice for farmers to send in the aborted foetus to the laboratory for tests. Unfortunately, in many situations this may not be very productive because abortions often result from a problem with the placenta or the womb, and the expulsion of the foetus is the end result rather than the cause. As a general rule, the placenta should be submitted as well if available, using the same approach with regard to packaging as for autopsy material. Large enough containers may not be available, however, in which case the material can be placed in a plastic bag and securely tied, making sure that there are no leaks.

2.7   Skin scrapings and biopsies

Diseases of the skin which produce scabby lesions, e.g. ringworm and mange, can usually be diagnosed by examination in the laboratory of scrapings of the lesions. Using a clean sharp blade such as a scalpel or razor blade, the best site to take a scraping is always the edge of the lesion where the causative organisms are likely to be most active and abundant. The scrapings should be taken into a clean dry container. Virtually anything will suffice so long as the scrapings can easily be emptied out in the laboratory, for example, small jars with a lid or even envelopes which can be sealed. Small cardboard packets are not suitable as the scrapings are liable to get trapped in the flaps or even be lost.

Some skin diseases produce nodular rather than scabby lesions, in which case biopsies of the actual nodules have to be sent to the laboratory if a diagnosis is required. Using a sterile sharp scalpel blade, a sample of a nodule can be cut out and handled in the same way as other autopsy material. Such a procedure will, of course, result in some bleeding and the site must be treated as if it were a wound as described above.

2.8   Faeces samples and rectal swabs

Samples of faeces are commonly examined in the laboratory for examination for eggs of internal parasites (helminths and coccidia, see Chapter 2). No special skill is required. Faeces can be collected by hand from the rectum taking care not to damage the rectal lining in the process. Disposable plastic gloves are often used, but this is not essential as long as the hands are thoroughly washed afterwards. The most suitable containers are small plastic pots with snap-top lids which hold about 10 g of faeces, but small clean jars with secure lids are just as effective. Small containers are most suitable as they should be as full as possible with the minimum of air, otherwise there is a greater risk of any worm eggs

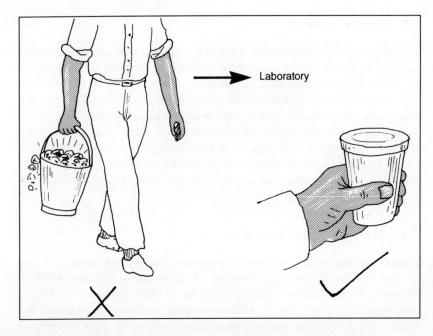

Fig 8.9   A 10 gram sample of faeces tightly packed in a sealed container should suffice for most laboratory tests on faeces.

hatching to larvae and affecting the worm egg counts in the laboratory. Because faeces are readily available and often abundant, farmers often tend to submit far more than is required. A sample of about 10g should suffice for most situations (Fig 8.9).

Faeces samples may also be used to isolate pathogenic organisms, such as causes of diarrhoea. In this case, however, it is usually better to submit a faecal swab taken from the rectum. This is more hygienic than samples of faeces and just as satisfactory from the laboratory point of view.

2.9   General points about packaging, labelling and history

As already mentioned, it is unlikely that farmers will have equipment, preservatives and proper containers for the collection and submission of samples to the laboratory, and so some sort of sensible compromise is usually necessary. With modern shopping practices today, an array of containers (tubs, jars, etc.) make their way into many households, and it is often possible to compromise and find suitable containers that can be cleaned and tightly secured. Whatever the container(s), the samples should be packed tight and not loose, using some sort of padding if necessary, to minimise the risk of breakage in transit. Most samples submitted by farmers contain no preservative, in which case they should be kept as cool as possible and submitted to the laboratory as quickly as possible. If any delay is likely, the samples should be transported on ice in a suitable container, such as in a flask for small samples, or in an insulated 'picnic-box' for larger samples.

NB   Unless the above can be carried out, unpreserved samples in the tropics quickly putrefy and become quite useless for laboratory tests.

Samples sent to the laboratory must be clearly labelled and accompanied by a history. The temptation to write the history on the sample container should be avoided. It is far better to write a simple code (e.g. a number or letter) on the container and refer to it in a written history on a separate sheet of paper. The history should contain as much information as possible about the disease in question, for example, which animals are affected, the clinical signs, how many are at risk, the timescale, any recent changes such as newly-introduced animals, change of grazing and any treatment given.

History writing is, to a certain extent, an art which only comes with a certain amount of practice. Sadly, even experienced veterinarians do not always master this particular art but providing farmers use common sense when submitting samples and include all the information they think is relevant to the particular disease episode in question, they won't go far wrong.

Glossary

Acarines Ticks and mites

Acaricide Drug that kills ticks and mites

Active or acquired immunity Immunity to disease following recovery from, or vaccination against that disease

Acute disease Disease that runs a short course of a few days

Anti-coagulant Chemical to prevent clotting of blood samples

Antibiotic Chemical product of micro-organisms that kills or inhibits other micro-organisms and is used to treat infectious diseases

Antibody Serum protein (immunoglobulin) produced by the immune system against an antigen

Antigen Any foreign organism, chemical or toxin, that stimulates the production of antibodies

Antiseptic Chemical agent which inhibits micro-organisms and is suitable for treatment of skin wounds, etc.

Antiserum or hyper-immune serum Serum with high levels of antibody obtained by repeat exposure to an antigen; used for treatment of, or temporary protection against that antigen

Attenuated vaccine Live vaccine in which the antigen's virulence is reduced by passaging it through a series of live animals or cultures

Autonomic nervous system Part of the central nervous system that controls functions of which the animal is unaware

Bacillus Rod-shaped bacteria

Bacteria Single-celled rigid walled micro-organism. Many species cause infectious diseases

Bilirubin Bile pigment formed from blood haemoglobin

Broad spectrum antibiotic Antibiotic effective against a broad range of bacterial infections

Bronchus Tubes branching from the trachea into the lungs

Case mortality rate Proportion of animals affected with a disease to die of that disease

Cellular immunity Immunity in which T lymphocytes and other cells stimulated by them inhibit antigens

Cestodes See tapeworms

Chemotherapeutic Synthetic chemical which kills or inhibits micro-organisms

Chemoprophylaxis See prophylaxis

Chronic disease Disease that runs a long course of at least several weeks, but can be months or years

Chlamydiae Small intra-cellular bacteria

Clinical signs Detectable change in tissue structure or function as a result of disease

Coccus Round-shaped bacteria

Coitus The act of sexual intercourse

Colostrum First milk after parturition; rich in antibodies

Condition scoring Technique to assess an animal's body condition

Congenital disease and/or infection Disease or infection of the newborn acquired in the womb during pregnancy

Conjunctiva Mucous membrane lining the eyeball and the lining of the eyelids

Cyanosis Blue colouration of visible mucous membranes due to lack of oxygen from the circulation

Cyclical transmission Vector transmission of an infectious organism in which the organism develops in the vector

Dysentery Passage of faeces or diarrhoea containing blood

Disinfectant Chemical agent that inhibits or destroys micro-organisms; used to clean inanimate objects

Dyspnoea Difficult or laboured breathing

Ecchymoses Small haemorrhages in skin or mucous membranes

Encephalitis Inflammation of the brain

Encephalomyelitis Inflammation of the brain and spinal cord

Erythrocytes Red blood cells which transport oxygen to tissues

Flatworms Flat-shaped helminths (flukes and tapeworms)

Flukes Leaf-shaped flatworms; scientific name is trematodes

Flutter valve Valve used to ensure smooth delivery by gravity of large volumes of intra-venous infusions

Fomite Inanimate object which can convey disease agents

Fungi Spore-forming organisms widespread in nature; some species infectious and pathogenic to livestock

Gastro-enteritis Inflammation of the stomach and intestines

Gestation period Pregnancy period from conception to parturition

Haemoglobin Pigmented oxygen-carrying protein in erythrocytes

Haemoglobinuria or 'redwater' Condition in which free haemoglobin from destroyed erythrocytes is excreted in the urine

Haemolysis Destruction of erythrocytes

Hard ticks Ticks which have a hard protective cover

Helminths Parasitic worms

Humoral immunity Immunity in which B lymphocytes transform to plasma cells to produce antibodies

Hyper-immune serum See antiserum

Hypocalcaemia Below normal level of calcium in the blood

Icterus   See jaundice

Immunoglobulins   See antibodies

Injections   Inoculations by syringe of medicines into either the muscle, veins or under the skin

Intracellular micro-organisms   Micro-organisms that invade cells of the infected host, e.g. viruses, chlamydiae, rickettsiae and certain protozoa

Jaundice   Condition in which excessive production of bilirubin (see above) stains tissues yellow. The scientific term for jaundice is icterus

Leucocytes   White blood cells

Lymph node   See lymphatic system

Lymph vessel   See lymphatic system

Lymphatic system   Circulation of immune cells (lymphocytes) through a network of connecting lymph and blood vessels and accumulations of lymphoid tissues (lymph nodes)

Lymphocyte   Type of leucocyte in lymphatic system

Macrophages   Phagocytosing tissue cells derived from monocytes, a type of leucocyte (see phagocytosis below)

Mastitis   Inflammation of the mammary gland

Mechanical transmission   Simple, direct transfer of an infectious organism by a vector with no development of the organism in the vector

Micro-organism   Organisms only detectable under the microscope

Monocyte   Leucocyte that forms a macrophage in tissues

Morbidity rate   Proportion of a population affected by a disease

Mucous membrane   Lining of certain tubular and hollow organs which secretes mucus (see visible mucous membranes)

Mucus   Slime secreted by mucous membranes

Mycotoxicoses   Poisoning from certain toxin producing fungi

Myiasis   Invasion of tissues of live animals by fly larvae

Nematodes   See roundworms

Neutrophil   Phagocytosing leucocyte (see phagocytosis below)

Oestrus   Time when sexually active female accepts the male

Oestrus cycle   Time between one oestrus period and the next

Oribatid mites   Small free living mites found in pastures and soil which are intermediate hosts for certain tapeworms

Passive immunity   Temporary immunity from transfer of antibodies by inoculation of antiserum or sucking colostrum

Pathogen   Any agent that causes disease

Pediculosis   Infestation with lice

Peracute disease   Very rapid disease lasting up to one or two days

Petechiae   Pin-prick haemorrhages in skin or mucous membranes

Phagocytosis   Process of engulfing foreign organisms or particles by certain cells, e.g. macrophages

Placenta   Vascular protecting and nourishing membrane connecting the embryo to the mother in the womb

Plasma cell   Antibody-forming cells derived from B lymphocytes

Population mortality rate   Proportion of a population to die of a disease

Predisposing factor   Factor that increases the chance of disease

Prophylactic drug   Long acting drug used to prevent disease

Prophylaxis   Prevention of disease, e.g. if by use of drugs it is called chemoprophylaxis

Protozoa   Highest form of micro-organism; possess a nucleus

Respiration rate   Number of breathing cycles per minute

Rickettsiae   Small intra-cellular bacteria usually transmitted by arthropods

Roundworms   Round helminths; scientific name is nematodes

Scolex   Head of a tapeworm

Sensory nervous system   Part of the central nervous system which the animal controls, e.g. for walking, seeing, etc.

Septicaemia   Presence of micro-organisms and their toxins in the blood

Soft ticks   Types of ticks that do not possess a hard shell

Spirochaetes   Spiral shaped bacteria

Stillbirth   Birth of a dead fully-formed foetus

Subacute disease   Disease lasting about a week or so

Subclinical   Infection that does not produce clinical signs

Tapeworms   Flatworms comprised of a scolex and segments; scientific name is cestodes

Toxaemia   Presence of toxins in the blood

Toxin   Poison produced by living organisms

Toxoid   Vaccine based on an inactivated toxin

Trachea   The windpipe

Trematodes   See flukes

Vaccine   Antigen processed so that it can be administered to stimulate immunity without causing disease

Vector   Agent that transmits an infectious micro-organism from one animal to another, e.g. ticks

Venereal infection   Disease spread by coitus

Vertical transmission   Transfer of a disease from one generation to the next, e.g. from dam to embryo in pregnancy

Vibrios   Curved shaped bacteria

Viruses   Smallest micro-organisms; can only replicate inside cells of other organisms; many species are infectious and cause disease

Visible mucous membrane   Parts of mucous membranes that can be seen, e.g. the conjunctiva in the eye and the gums

Bibliography

Andrews, A. H., with Blowey, R. W., Boyd, H. and Eddy, R. G. (1992), *Bovine Medicine*, Blackwell Scientific Publications, Oxford, UK.

Blood, D. C. and Radostits, O. M. (1989), *Veterinary Medicine*, 7th edn, Baillière Tindall, London, UK.

Brander, G. C., Pugh D. M., Bywater, R. J. and Jenkins, W. L. (1991), *Veterinary Applied Pharmacology and Therapeutics*, 5th edn, Baillière Tindall, London, UK.

Carter, G. R. (1986), *Essentials of Veterinary Bacteriology and Mycology*, 3rd edn, Lea and Febiger, Philadelphia, USA.

Cockrill, W. R. (1974), *The husbandry and health of the domestic buffalo*, FAO, Rome, Italy.

FAO (1984), *Ticks and tick-borne disease control – a practical field manual (Volumes I and II)*, FAO, Rome, Italy.

Gatenby, R. M. (1991), *Sheep – The Tropical Agriculturalist*, Macmillan Education Ltd., London and Basingstoke, UK.

Geering, W. A. and Forman, A. J. (1987), *Animal Health in Australia, Volume 9, Exotic Diseases*, Australian Government Publishing Service, Canberra, Australia.

Henderson, D. C. (1990), *The Veterinary Book for Sheep Farmers*, Farming Press, Ipswich, UK.

Higgins, A. (1986), *The camel in health and disease*, Baillière Tindall, London, UK.

Leese, A. S. (1927), *A Treatise on the One-humped Camel in Health and in Disease*, Haynes and Son, Stamford, UK.

Lefèvre, P-C. (1991), *Atlas des maladies infectieuses des ruminants*, Le Centre Technique de Cooperation Agricole et Rurale (CTA) et Agence de Cooperation Culturelle et Technique.

Merck Veterinary Manual, 7th edn (1991), Merck and Co., Inc., Rahway, N.J., USA.

Ministry of Agriculture Fisheries and Food (1984), *Manual of Veterinary Investigation Laboratory Techniques, Volumes 1 and 2*, HMSO, London, UK.

Nicholson, M. J. and Butterworth, M. H. (1985), *A Guide to Condition Scoring Of Zebu Cattle*, International Livestock Centre for Africa, Addis Ababa, Ethiopia.

Payne, W. J. A. (1990), *An Introduction to Animal Husbandry in the Tropics*, 4th edn, Longman Scientific and Technical, Harlow, UK.

Roitt, I. M. (1988), *Essential Immunology*, 6th edn, Blackwell Scientific Publications, Oxford, UK.

Sewell, M. M. H. and Brocklesby, D. W. (1990), *Handbook on Animal Diseases in the Tropics*, 4th edn, Baillière Tindall, London, UK.

Urquhart, G. M., Armour, J., Duncan, J. L., Dunn, A. M. and Jennings, F. W. (1987), *Veterinary Parasitology*, Longman Scientific and Technical, Harlow, UK.

West, G. (1992), *Black's Veterinary Dictionary*, A. and C. Black, London, UK.

Wilson, R. T. (1984), *The Camel*, Longman Group Ltd., Harlow, UK.

Index

(Page numbers in **bold** refer to illustrations)

162

abnormal gait 87–8
diarrhoea in 81
killing diseases of 46–8
loss of condition diseases 52–3
mucosal discharges 69–70
nervous diseases 87–8
reproductive diseases 94–5
respiratory diseases 66
with scrapie **54**
skin diseases 58–9
visible mucous membranes, changes in 77
sheep pox 4, 47, 53, 59
shelter: and disease 124–6
temporary 125–6
shipping fever 65, 68
skin: biopsies 153
condition of 36–7
infection through 3–4
irritation 53
lesions 53–62
lumps and nodules 54–5
scrapings 153
snails, aquatic: and liver fluke 22
sodium deficiency 51, 86
staphylococcal mastitis 32
stephanofilarosis 57
Stilesia hepatica 20
stillbirths 92, 93–6
Strongyloides worms 17, 18
strongylosis 50
struck 48
sub-clinical infections 11–12
summer sores 61
sweating sickness 57, 68
sweet itch 61, 62
syringes and needles 128, 139–40
disposable 128
sizes of 139–40

Taenia saginata cysts **20**
Talfan disease 89
tapeworms 20–1, 22
and mites 15
Teschen disease 89
tetanus 50, 88, 90
theileriosis 11, 31, 33, 46, 48, 72, 76, 77, 86
bovine **151**
tickborne 119
thornyheaded worms 19
tickborne: diseases 117–21
fever 95

ticks 14, 15
as disease agents 14, 16, 27, 31
infestation 51
paralysis 86, 97
tissue: and helminths 23, 24
torsalo 57
toxaemia: pregnancy 47
toxins 8
from bacteria 28–9
see also exotoxins
Toxocara vitulorum 19
toxoids 133
transmission: of disease 2, 16, 31
Trichomonas foetus 31
trichomonosis 31, 93
Trypanosoma evansi 129
Trypanosoma simiae 49
trypanosomoses 31, 90
fly-borne infections 111–12
trypanosomosis 10, 11, 16, 32, 50, 51, 53, 61, 71, 76, 77, 97
tsetse fly 10
tuberculosis 51, 53, 65, 66, 68, 97

ulcer, lymphangitis 61
urticaria 57, 61, 62

vaccinations 130–4
vaccines 131–4
Vacutainer® equipment **148, 149**
vectors, disease 5, 16
venereal infectious diseases 1, 4, 12
distribution 99, 109
vertical transmission: of disease 4
veterinary principles: general 123–38
veterinary procedures: general 139–55
virulence: of viruses 26
viruses 5, 25–6
visible mucous membranes: see under mucous membranes
visna 53, 88

warble fly 57
weak newborn 94–5
worm nodule disease 57, 61, 97
worms: see helminths
wounds: care of 128–9
skin, treatment 145–6

yeasts 30